Making the Soldier Decisive on Future Battlefields

Committee on Making the Soldier Decisive on Future Battlefields

Board on Army Science and Technology

Division on Engineering and Physical Sciences

NATIONAL RESEARCH COUNCIL
OF THE NATIONAL ACADEMIES

THE NATIONAL ACADEMIES PRESS
Washington, D.C.
www.nap.edu

THE NATIONAL ACADEMIES PRESS 500 Fifth Street, NW Washington, DC 20001

NOTICE: The project that is the subject of this report was approved by the Governing Board of the National Research Council, whose members are drawn from the councils of the National Academy of Sciences, the National Academy of Engineering, and the Institute of Medicine. The members of the committee responsible for the report were chosen for their special competences and with regard for appropriate balance.

This study was supported by Contract No. W911-NF-11-C-0099 between the National Academy of Sciences and the Department of Defense. Any opinions, findings, conclusions, or recommendations expressed in this publication are those of the author(s) and do not necessarily reflect the views of the organizations or agencies that provided support for the project.

Front cover photograph courtesy of U.S. Army

International Standard Book Number -13: 978-0-309-28453-0
International Standard Book Number -10: 0-309-28453-8

Limited copies of this report are available from:

Board on Army Science and Technology
National Research Council
500 Fifth Street, NW, Room 940
Washington, DC 20001
(202) 334-3118

Additional copies are available from:

The National Academies Press
500 Fifth Street, NW
Keck 360
Washington, DC 20001
(800) 624-6242 or (202) 334-3313
http://www.nap.edu

Printed in the United States of America

COMMITTEE ON MAKING THE SOLDIER DECISIVE ON FUTURE BATTLEFIELDS

HENRY J. HATCH, NAE,[1] LTG, U.S. Army (retired), *Chair*, Independent Consultant, Oakton, Virginia
W. PETER CHERRY, NAE, Independent Consultant, Ann Arbor, Michigan
PAUL W. GLIMCHER, New York University Center for Neural Science
RANDALL W. HILL, JR., University of Southern California Institute for Creative Technologies, Marina del Rey
ROBIN L. KEESEE, Independent Consultant, Fairfax, Virginia
ELLIOT D. KIEFF, NAS/IOM, Channing Laboratory, Harvard University, Boston, Massachusetts
JEAN MACMILLAN, Aptima, Inc., Woburn, Massachusetts
WILLIAM L. MELVIN, Georgia Tech Research Institute, Smyrna
RICHARD R. PAUL, Maj. Gen. U.S. Air Force (retired), Independent Consultant, Bellevue, Washington
RICHARD PEW, BBN Technologies, Cambridge, Massachusetts
M. FRANK ROSE, Radiance Technologies, Huntsville, Alabama
ALBERT A. SCIARRETTA, CNS Technologies, Springfield, Virginia
ANN SPEED, Sandia National Laboratories, Albuquerque, New Mexico
JOSEPH YAKOVAC, LTG, U.S. Army (retired), JVM LLC, Hampton, Virginia

Staff

BRUCE A. BRAUN, Director, Board on Army Science and Technology
ROBERT LOVE, Study Director
NIA D. JOHNSON, Senior Research Associate, Board on Army Science and Technology
DEANNA SPARGER, Program Administrative Coordinator, Board on Army Science and Technology
JOSEPH PALMER, Senior Program/Project Assistant

---

[1]NAE, National Academy of Engineering; IOM, Institute of Medicine; NAS, National Academy of Sciences

Preface

This study resulted from recognition by the U.S. Army that a great disparity exists between the decisive overmatch capability, relative to prospective adversaries, of major U.S. weapon systems (such as tanks, fighter aircraft, or nuclear submarines) and the relative vulnerability of dismounted soldiers when they are operating in small, detached units (squads). The increased reliance in recent Army deployments on soldiers operating in these tactical small units (TSUs), as well as the expanding responsibilities of ground forces in the future for missions that go beyond traditional combat, provided the incentive to ask what could be done to give dismounted soldiers and TSUs a credible degree of decisive overmatch in any of the anticipated future operational environments.

I would like to thank the Committee on Making the Soldier Decisive on Future Battlefields for its tenacity and dedication in interviewing numerous experts (including recently deployed Army enlisted soldiers and officers), assessing the pertinent issues, and developing recommendations to address the many demands of its statement of task from the Army sponsor (see Summary of this report). The committee in turn is grateful to the many Army and Department of Defense personnel, both civilian and military, who provided much of the information on which this report is based. We particularly thank the veterans of recent combat deployments who shared with us their hopes for those who will follow them, as well as their insights, frustrations, and triumphs in the trying circumstances of operations in Iraq and Afghanistan.

The committee and I also greatly appreciate the support and assistance of the National Research Council (NRC) staff, which ably assisted the committee in its fact-finding activities and in production of the report. In particular, I thank the staff of the NRC's Board on Army Science and Technology (BAST), who successfully organized the attendance of committee members and guests at major meetings in multiple locations and maintained a secure Internet forum for the members to accumulate study information, collaborate on report inputs, share expertise, and develop the consensus for the report we present here.

The study was conducted under the auspices of the BAST, a unit of the NRC's Division on Engineering and Physical Sciences, established in 1982 at the request of the United States Army. The BAST brings broad military, industrial, and academic scientific, engineering, and management expertise to bear on technical challenges of importance to senior Army leaders. The BAST is not a study committee; rather, it deliberates on study concepts and statements of task for the expert committees that are formed under rigorous NRC procedures to conduct a particular study. The BAST discusses potential study topics and tasks, ensures study project planning and execution in conformance with NRC procedures, and suggests candidate experts to serve as committee members or report reviewers.

Although the Board members are listed in the front pages of the report, with the exception of any members who were nominated and appointed to the study committee, they were not asked to endorse the committee's findings or recommendations or to review final drafts of the report before its release. The findings and recommendations are those reached by unanimous consensus of the Committee on Making the Soldier Decisive on Future Battlefields. The NRC's approval of this report likewise does not indicate a position on the substance of the findings and recommendations but rather certifies that the study was conducted in accordance with its procedures.

Hank Hatch, *Chair*
Committee on Making the Soldier Decisive
on Future Battlefields

Acknowledgments

This report has been reviewed in draft form by individuals chosen for their diverse perspectives and technical expertise, in accordance with procedures approved by the National Research Council's Report Review Committee. The purpose of this independent review is to provide candid and critical comments that will assist the institution in making its published report as sound as possible and to ensure that the report meets institutional standards for objectivity, evidence, and responsiveness to the study charge. The review comments and draft manuscript remain confidential to protect the integrity of the deliberative process. We wish to thank the following individuals for their review of this report:

Ruzena K. Bajcsy (NAE/IOM), University of California, Berkeley
Nancy J. Cooke, Arizona State University
Gilbert F. Decker, Consultant
William P. Delaney (NAE), MIT Lincoln Laboratory
Richard Dempsey, U.S. Army
Charles B. Duke (NAE/NAS), Xerox Corporation (retired)
Harry W. Jenkins, U.S. Marine Corps (retired)
Roger L. McCarthy (NAE), McCarthy Engineering
Stewart D. Personick (NAE), New Jersey Institute of Technology
Dennis J. Reimer, U.S. Army (retired)
Robert H. Scales, Colgen, Inc.
Daniel P. Siewiorek (NAE), Carnegie Mellon University
Judith L. Swain (IOM), National University of Singapore
Michael R. Thompson, Scitor Corporation

Although the reviewers listed above have provided many constructive comments and suggestions, they were not asked to endorse the conclusions or recommendations, nor did they see the final draft of the report before its release. The review of this report was overseen by Robert A. Frosch, NAE, Harvard University. Appointed by the National Research Council, he was responsible for making certain that an independent examination of this report was carried out in accordance with institutional procedures and that all review comments were carefully considered. Responsibility for the final content of this report rests entirely with the authoring committee and the institution.

Contents

SUMMARY 1

1 INTRODUCTION 15
Origin of the Study, 15
Areas of Focus, 15
Definitions, 17
Study Approach, 18
Essential Principles to Achieve TSU Overmatch, 19
Committee's Approach to the Human Dimension, 20
Report Organization, 22
References, 24

2 CAPABILITIES 25
TSU Missions and Tasks, 27
Situational Understanding, 31
The Role of Decision-Making in Overmatch, 31
Three Levels of Situational Awareness, 32
Network Integration, 33
Military Effects, 36
Lethality, 36
Stability and Humanitarian Effects, 38
Maneuverability, 40
Sustainability, 41
Power and Energy, 41
Survivability, 43
Individual Soldier Protection, 44
TSU Protection, 45
Layers of Protection External to the TSU, 46
Current Operational Weaknesses, 46
Human Dimension Issues, 47
Deficits in TSU and Soldier Training, 51
Deficits in the Analytical Foundation for Building Decisive TSUs, 54
The Untapped Human Dimension, 55
Prospective Solution Categories, 55
References, 56

3 SETTING THE CONDITIONS TO ACHIEVE SOLDIER AND TSU OVERMATCH 59
Placing Emphasis on the Human Dimension, 60
Systems Engineering For Decisive Overmatch, 62

Metrics for the Dismounted TSU and Soldier, 66
Streamlining Acquisition of Solutions to Achieve TSU Overmatch Capabilities, 69
References, 73

4 ACHIEVING OVERMATCH 75
Designing the TSU, 76
TSU Design Considerations, 77
Soldier Performance, 78
Soldier Selection, 82
Focusing on TSU Training, 85
The TSU Training Imperative, 85
Findings and Recommendations on TSU Training, 91
Integrating the TSU into Army Networks, 92
Definition of Network Integration, 93
Potential Benefits, 94
DOTMLPF Considerations, 96
Network Integration Priorities, 100
Balancing TSU Maneuverability, Military Effects, and Survivability, 105
Soldiers Carry Too Much to Move Quickly, Act Effectively, and Avoid Injury, 106
Potential Benefits of Optimizing TSU and Soldier Systems for Maneuverability, Military Effects, and Survivability, 108
Selected DOTMLPF Opportunities for Balancing Maneuverability, Military Effects, and Survivability, 112
Findings and Recommendations for Achieving TSU Balance, 118
Leveraging Advances in Portable Power, 121
DOTMLPF Considerations, 121
Battery and Fueled Energy Storage Systems, 122
Energy Harvesting, 125
References, 127

APPENDIXES

A Biographical Sketches of Committee Members 131
B Committee Meetings 139
C Army Terminology and Doctrine Relevant to Dismounted Soldier Missions 143
D History and Status of Design for the Soldier as a System 157
E Measures of Performance and Measures of Effectiveness 165
F Simulation Technologies and Devices 171
G Technology Solutions for TSU Sensor Missions 175
H Prospective Robotics Technologies 197
I Energy Technologies and Applications for the Soldier 207
J Lethal and Nonlethal Weapons 227

Tables, Figures, and Boxes

TABLES

2-1 Energy Formats and Amounts of Energy Required for a 72-Hour Mission, 43

G-1 Squad-Level Sensor Considerations, 179
G-2 Squad-Level Sensor Missions, 182
G-3 Situational Awareness Sensor Technology Gap Assessment, 188
G-4 Force Protection Sensor Technology Gap Assessment, 191
G-5 Precision Targeting Gap Assessment, 194

I-1 State of the Art for Technologies Most Relevant to the Dismounted Soldier, 208
I-2 Fuel-cell State of the Art, 212
I-3 Near-term State of the Art for Relevant Combustion Technologies, 216
I-4 Energy and Weight of Biomechanical Prototype, 223
I-5 State of the Art in Hybrid Systems, 224

FIGURES

2-1 Decreasing size of fighting unit with critical influence and increasing area of operation for a tactical small unit, 29
2-2 Generic U.S. Army rifle squad, 38
2-3 The modern warrior with combat load during dismounted operation in Afghanistan, 42

4-1 Notional information requirements for TSU with offensive mission, 97
4-2 Soldier power solutions, 123
4-3 Comparison of energy options for the dismounted Soldier, 124

C-1 The elements of combat power, 151

D-1 Soldier with combat load, 160

H-1 Protection Ensemble Test Mannequin (PETMAN) robot without external shell, 199
H-2 BigDog, 201
H-3 LS3, Legged Squad Support System, 202

I-1 Selected energy densities, 207
I-2 Deployable portable solar array, 219
I-3 Harvest of biomechanical energy, 221
I-4 Soldier Power Regeneration Kit, 222

J-1 Major improvements to individual and crew served weapons over the last 10 years, 229
J-2 Current and future crew served weapons, 230
J-3 Enhanced sniper technologies, 231
J-4 Sensors and lasers functions, 232
J-5 Distribution of mortars in Army units, 234
J-6 Examples M224A1 Lt Wt 60mm mortars, 235
J-7 MI50/M151 MFCS dismount, 236
J-8 Accelerated precision mortar, 237

BOXES

SUM-1 Statement of Task, 2

1-1 Statement of Task, 16

3-1 Military Implementation of Human-Systems Integration, 65

Acronyms and Abbreviations

AAR	after action review
AIT	advanced individual training
ACEP	Army Center for Enhanced Performance
APFT	Army Physical Fitness Test
APS	active protection systems
AO	area of operation
ARDEC	Armament Research, Development and Engineering Center
ARFORGEN	Army Force Generation
ARI	Army Research Institute for the Behavioral and Social Sciences
ARL	Army Research Laboratory
ARL-HRED	Army Research Laboratory-Human Research and Engineering Directorate
ARTEMIS	All-Terrain Radar for Tactical Exploitation of MTI and Imaging Surveillance
ASA(ALT)	Assistant Secretary of the Army (Acquisition, Logistics and Technology)
ASB	Army Science Board
ASIMO	Advanced Step in Innovative Mobility
ASVAB	Armed Services Vocational Aptitude Battery
BAST	Board on Army Science and Technology
BCT	brigade combat team
BT	basic training
C4ISR	Command, Control, Communications, Computers, Intelligence, Surveillance, Reconnaissance
CALL	Center for Army Lessons Learned
CASCOM	U.S. Army Combined Arms Support Command
CBA	cost benefit analysis
CBRN	chemical, biological, radiological, and nuclear
CCD	capabilities development document
CERDEC	Communications-Electronics Research, Development, and Engineering Center
CIED	Counter-improvised explosive device
CIST	company intelligence support team
CONOPS	concept of operations
CRAM	Counterrocket, artillery, and mortar
DA	Department of the Army
DARPA	Defense Advanced Research Projects Agency
DIME	diplomatic, information, military, and economic
DMFC	direct methanol fuel cell
DoD	Department of Defense

DOTMLPF	Doctrine, Organization, Training, Materiel, Leadership and Education, Personnel and Facilities
DoS	U.S. Department of State
DSTS	Dismounted Soldier Training System
EEA	essential element of analysis
FCS	future combat systems
FOB	forward operating base
FORESTER	FOliage PENetration Reconnaissance, Surveillance, Tracking and Engagement Radar
FITE	Future Immersive Training Environment
FMV	full motion video
G1	Deputy Chief of Staff for Personnel
GMAV	gas micro air vehicle
GPS	global positioning system
GSM	global system for mobile communication
HAL	Hybrid Assistive Limb
HHC	headquarters and headquarters company
HSI	Human-Systems Integration
HSI/MSI	hyperspectral imaging/multispectral imaging
HULC	human universal load carrier
ICD	initial capabilities document
IED	improvised explosive device
IIT	Infantry Immersion Trainer
IMPRINT	Improved Performance Research Integration Tool
IPE	individual protective equipment
IR	infrared
IRST	infra-red search and track
IRT	independent review team
ISR	intelligence, surveillance, and reconnaissance
JIEDDO	Joint Improvised Explosive Device Defeat Organization
JCIDS	Joint Capabilities Integration and Development System
JCTD	Joint Capabilities Technology Demonstration
JP	jet propellant
LADAR	LAser Detection And Ranging
L-V-C	Live - Virtual - Constructive
MANPRINT	MANpower, PeRsonnel, INTegration
MCoE	Maneuver Center of Excellence
MEPS	Military Entrance Processing Stations

METT-TC	Mission, Enemy, Terrain and weather, Troops and support available—Time available, Civilians
MMOG	massively multiplayer online game
mmw	millimeter wave
MOE	measure of effectiveness
MOP	measures of performance
MOS	military occupational specialty
MOUT	military operations on urban terrain
MRMC	U.S. Army Medical Research and Materiel Command
NCO	non-commissioned officer
NGO	nongovernmental organization
NPC	non-player characters
NRC	National Research Council
NSWC	Naval Surface Warfare Center
OEF	Operation Enduring Freedom
OE/OD	organizational effectiveness/organizational development
OIF	Operation Iraqi Freedom
OPTEMPO	operations tempo
ORSA	operations research and system analyst
OSA	open system architecture
OSUT	one-station unit training
OSHA	Occupational Safety and Health Administration
PEO Soldier	Program Executive Office-Soldier
PEO STRI	Program Executive Office for Simulation, Training and Instrumentation
PETMAN	protection ensemble test mannequin
POI	program of instruction
PMESI	political, military, economic, social, infrastructure
PSM	physiological status monitor
R&D	research and development
RDEC	U.S. Army Research, Development & Engineering Center
RMFC	reformed methanol fuel cell
ROEs	rules of engagement
RoL	Rule-of-Law
SA	situational awareness
SALTI	DARPA Synthetic Aperture Ladar for Tactical Imaging
SAR	search and rescue
SIGINT	signals intelligence
SIPRNet	Secure Internet Protocol Network
SOFC	solid oxide fuel cell
SOS	system-of-systems

SOT	Statement of Task
SSIM	strategic social interaction module
S&T	science and technology
STRICOM	Simulation Training and Instrumentation Command
SWAP	size, weight, and power
SWAP-C	size, weight, power, and cost
TAPAS	Tailored Adaptive Personality Assessment System
TCPED	tasking, collection, processing, exploitation and dissemination
TiGRNET	Tactical Ground Reporting Network
TOPS	Tier One Performance Screen
TRL	technology readiness level
TRADOC	U.S. Army Training and Doctrine Command
TSU	tactical small unit
TTHS	trainees, transients, holdees, and students
TTP	tactics, techniques and procedures
TUS	U.S. Navy Transparent Urban Structures
UAS	unmanned aerial system
UAV	unmanned aerial vehicle
UGV	unmanned ground vehicle
USAREC	United States Army Recruiting Command
USARIEM	U.S. Army Research Institute of Environmental Medicine
VBS2	Virtual Battle Space 2
VHF	very high frequency
WAMI	wide area motion imagery
WAS	wide area security
WLR	weapons location radar

Summary

The focus of this study was achieving decisive overmatch for the dismounted Soldier[1] and tactical small unit (TSU) on future battlefields. In particular, the committee was asked to determine the elements of overmatch capabilities necessary to achieve decisiveness, identify technical requirements for optimizing Soldiers and small units, identify near-, mid-, and far-term technologies for investment, and determine the relative importance of such investments. The complete statement of task is in Box SUM-1.

To examine the desired elements of decisive overmatch, Chapter 2 identifies capabilities needed for decisive overmatch by Soldiers and small units in dismounted infantry squad operations, including situational understanding, military effects (including lethal and nonlethal effects and stability actions), maneuverability, sustainability, and survivability. Chapter 3 then articulates the foundational capabilities needed to identify and implement potential solutions. Finally, Chapter 4 describes how to achieve overmatch by focusing on the five areas most likely to enable improvements in Soldier and TSU decisiveness.

To identify relevant technical requirements, the committee gathered information about ongoing concept and technology development efforts both in and out of the Army with potential to contribute to decisive overmatch within the near (5 years), mid (5-10 years), and far (beyond 10 years) terms. The committee also interviewed Soldiers, both officer and enlisted, with recent combat experience in Iraq and Afghanistan to gain an understanding of known shortcomings.

SETTING CONDITIONS TO ACHIEVE OVERMATCH

If decisive overmatch is to be achieved and sustained in the future, it is essential that the Army identify the favorable asymmetries that can be exploited and the unfavorable asymmetries that must be mitigated. Without the artifacts of a holistic systems engineering process applied to

---

[1]For this report, the committee has chosen to follow the Army's policy since 2003 of capitalizing the word "soldier" when it refers to a soldier in the U.S. Army.

BOX SUM-1 Statement of Task

The U.S. military does not believe its soldiers, sailors, airmen, and marines should be engaged in combat with adversaries on a "level playing field". Our combat individuals enter engagements to win. To that end, this country has used its technical prowess and industrial capability to develop decisive weapons, weapons that over-match those of potential enemies, such as the M1A2 tank, the F-22 fighter, and the Seawolf attack submarine. The country is now engaged in what has been identified as an "era of persistent conflict" in which the most important weapon is the dismounted soldier operating in small units. More than for soldiers in Vietnam, Korea, and WWII, today's soldier must be prepared to contend with both regular and irregular adversaries. Results in Iraq and Afghanistan show that while the US soldier is a formidable fighter, his contemporary suite of equipment and support does not enjoy the same high degree of overmatch capability exhibited by large weapons platforms—yet it is the soldier who ultimately will play the decisive role in restoring stability.

A study is needed to establish the technical requirements for overmatch capability for dismounted soldiers operating individually or in small units. What technological and organizational capabilities are needed to make the dismounted soldier a decisive weapon? How can technology help those soldiers remain decisive on a changing, uncertain and complex future environment? The study will examine the applicability of systems engineering to soldiers and small units, as well as specific technology areas that are relevant to making soldiers decisive, particularly in conditions where we still take casualties today (movement to contact and chance encounters). Technology areas to be considered should include (but not be limited to) situational awareness, weapons, mobility, and protection, adaptation to battlefield environments (e.g., clothing, cooling), communications and networking, human dynamics (e.g., physical, cognitive, behavioral), and logistical support (e.g., medical aid, food, water, energy).

The NRC will establish an ad hoc study committee to examine these requirements. The committee will:

1. Determine the elements of overmatch capabilities necessary for a dismounted soldier to be a decisive weapon on the battlefield, Consider both the individual soldier as well as the soldier as part of a small (squad-size or smaller) unit.
2. Identify technical requirements for optimizing soldiers and small units to achieve overmatch capabilities on the battlefield. Consider technology and societal trends that may affect the balance between U.S. forces & adversaries both now and in future years.
3. Identify near-term, mid-term and far-term technologies in which new or enhanced S&T investments would facilitate the development of decisive soldier capabilities.
4. Determine the relative importance of such investments in making the soldier decisive on future battlefields.

the Soldier and TSU, a final assessment of overmatch opportunities is not possible.

As it came to understand the non-materiel side of TSU and Soldier capabilities, the committee decided that the greatest returns on Army investments for improvements in the near, mid, and far terms would be achieved by balancing the materiel aspects of technology developments with non-materiel aspects, found primarily in the human dimension. The Committee defined the human dimension to include all the attributes of the individual Soldier and of the collected Soldiers forming the TSU that impact performance of mission tasks. This differs from the Army's current perspective on the human dimension, which does not adequately include the complexities of individual Soldier tasks and human interactions within teams.

Knowledge in fields such as cognitive psychology, sociology, and neuro-economics, can provide many of the answers to questions surrounding the adequacy and potential of the individual Soldier and TSU. Applying such knowledge will require an expansion of resources devoted to human dimension research and technology development, as well as to small-unit organization and doctrine.

The study concluded that an essential principle for achieving overmatch capabilities is to recognize that integrating the human dimension with materiel advances is at the core of all TSU improvements. However, Army research and development has always been insufficiently resourced to provide the range of human-dimension opportunities and solutions that might provide overmatching performance.

Recommendation 1: To determine overmatch options for the TSU, the Army should provide sufficient resources for the full range of human-dimension opportunities and solutions that might provide overmatching performance.

Get Serious About Systems Engineering

A systems engineering methodology is essential to develop the relevant measures of performance and effectiveness, as well as supporting indicators, for the TSU. Such measures can be used to develop an integrated assessment methodology (and associated tools) that can evaluate both materiel and non-materiel impacts of prospective TSU enhancements.

Recommendation 2: The Army should establish a Systems Engineering executive authority to support a system-of-systems engineering environment that will be responsible for developing methodologies and analytical tools to evaluate and acquire total system solutions for the dismounted Soldier and TSU. This executive authority must have sufficient seniority, influence, and budget control to operate effectively across the entire Army acquisition community (including research and development, test, and evaluation) in executing its systems engineering mission.

Establish Metrics

Improvement is needed in many human-dimension areas at the Soldier and TSU levels, including leader development, situational understanding, cognitive performance and overload, physical performance, mental and physical resiliency, cultural understanding, human-system interfaces, and other areas with potential to contribute to decisive overmatch. Current measures of performance (MOPs) and measures of effectiveness (MOEs) are not adequate to assess these improvements. In the past, use of MOPs and MOEs and other elements of analysis have focused on platform-centric evaluations, using improvements in Soldier and small-unit performance and effectiveness as measures for the benefits of the platform or other materiel system being assessed, rather than as measures of Soldier and TSU capabilities.

A rigorous methodology and comprehensive set of MOEs and MOPs that represent the performance and effectiveness of a TSU, including the capabilities and limitations of all components—the Soldiers and materiel systems—and their interactions, would provide objective measures that can be directed at the entirety of the TSU ensemble.

Recommendation 3: The Army should develop, maintain, and evolve an optimal set of measures of performance (MOPs) and measures of effectiveness (MOEs) for assessing capability improvements for the dismounted Soldier and TSU by investing in an analysis architecture and infrastructure, including a comprehensive metrics development methodology that supports objective dialogue among combat and system developers, systems engineers, trainers, and program activities. The MOPs and MOEs, together with the guidance for using them, should be tested and validated for practical application and ease of use, as well as for accuracy as predictors and indicators of desired performance and effectiveness outcomes.

Streamline Acquisition Processes

Despite the advice of multiple review teams on the importance of a holistic approach to development, procurement, and support of Soldier capabilities, the Army is still equipping the dismounted Soldier through separate programs of record. Army acquisition essentially consists of providing for multiple independent pieces, rather than providing for integrated systems. The urgency to support the force in the field during current operations has led to a reliance on rapid equipment fielding, which has exacerbated this stove-piped approach.

The acquisition system has been relatively unresponsive to the needs of dismounted Soldiers when compared to large weapons and mobility systems. The goal of achieving overmatch capabilities cannot be accomplished until small-unit and Soldier requirements are accorded the same high levels of attention as major materiel systems requirements. At the same time, the approach of acquiring and fielding every "new" technology is both impractical and unaffordable. Most important, it is unlikely that the solutions to achieve overmatch capabilities can be successfully implemented within the Army's current acquisition framework. A principled groundwork for analyzing the TSU system has not been laid for a natural progression to define and implement overmatch capabilities that integrate the span of human and materiel dimensions and that evolve continuously with changing threats and opportunities.

Recommendation 4: The Army should establish an executive authority for TSU integration, responsible for option generation and evaluation, requirements currency, and programmatic acquisition for the Soldier and TSU within a metrics-driven, system-of-systems engineering environment.

COMPONENTS OF CAPABILITY SOLUTIONS MOST LIKELY TO ACHIEVE OVERMATCH

The committee identified many opportunities to improve the capability of TSUs in ways that could potentially ensure that future TSUs have decisive overmatch across the range of military operations expected in future deployments. In the committee's judgment, many of these opportunities will have their greatest effect *only if* both materiel and non-materiel factors from across the DOTMLPF (Doctrine, Organization, Training, Materiel, Leadership and Education, Personnel and Facilities) domains are integrated in an optimized capability solution, in accordance with the four recommendations stated above. All these opportunities, or *capability options*, will have interactive consequences, positive and negative, that will require the rigorous assessment and design approach embodied in Recommendations 1 through 4 to find the optimal set. For this reason these four recommendations have higher priority for achieving TSU overmatch than the other eleven report recommendations.

The committee concluded that the capability solutions with the highest potential to contribute to decisive overmatch would likely fall into one or more of five solution areas:

- Designing the TSU
- Focusing on TSU Training
- Integrating the TSU into Army Networks
- Balancing TSU Maneuverability, Military Effects, and Survivability
- Leveraging Advances in Portable Power

Designing the TSU

The principles for achieving overmatch reflected in Recommendations 1 through 4 will allow the Army to leverage Soldier performance as never before and to determine the TSU design that will be dominant across the full range of combat and stability operations. A systems approach that focuses on developing TSU metrics can expand TSU design options, enabling the Army to fully exploit the capabilities of Soldiers and equipment. The TSU should not be viewed as just an organization or formation but as a system of systems. A holistic, top-down analysis would then be able to determine design parameters for the optimal size (number of Soldiers), organization (number of fire teams, duties), and equipment (communication, lethality systems, etc.) of the dismounted TSU of the future.

Development and analyses of TSU options will require collaboration among multiple Army activities, including the U.S. Army Training and Doctrine Command (TRADOC) Infantry School at the Maneuver Center of Excellence (MCoE), the TRADOC Analysis Center, the Army Research Laboratory , the Army Research Institute of Environmental Medicine, the Army Research Institute for the Behavioral and Social Sciences, the Army Materiel Systems Analysis Activity, and the Army program executive offices for Simulation, Training, and Instrumentation(PEO STRI) and Soldier (PEO Soldier).

Recommendation 5: The Army should transform and sustain the design of the TSU, including re-assessing unit organization and size, by the following actions:

a. Develop representative measures of performance (MOPs) and measures of effectiveness (MOEs) for the primary dimensions of TSU performance, and ensure these measures incorporate human dimension criteria.

b. Assemble a consortium of stakeholders to implement iterative work-centered analyses of the Soldier task workload and the TSU and Soldier-system performance required by increasing the scope (range, quality, thresholds) of TSU MOPs MOEs. The analyses should enable development of predictive analytical models of Soldier physical and cognitive task and mobility performance, Soldier-to-Soldier task and mobility interaction within a TSU network, and TSU task and mobility performance.

c. Expand the TSU task and mobility model to predict influences of weapons, intelligence, surveillance, and reconnaissance, and information technologies on TSU MOPs and MOEs.

Changes in TSU design will require not only considerations for future missions and equipment but also adequate attention to the Soldier as a human. Capabilities of the TSU and of the Soldiers in it are highly dependent on each other. Enhancements to TSU performance and effectiveness should also enhance performance and effectiveness of the individual Soldier. Likewise, Soldier enhancements should increase the performance and effectiveness of the TSU. For example, situational awareness within the TSU enhances an individual Soldier's situational awareness. Enhancing the shooting skill of one Soldier will, in turn enhance the lethality of the TSU. Future enhancements to the TSU and Soldiers should be designed to provide a synergistic effect that is greater than the sum of incremental improvement from each enhancement by itself.

Achieving decisive Soldier performance requires several near-term actions. These include the following.

- Institutionalizing the functions of the Army Center for Enhanced Performance;

- Assembling a "Physiological Readiness Check List" for use in training and operational testing and refining development of nonintrusive physiological status monitors;
- Expanding research in the social processes of small units; and,
- Expanding research in individual differences, especially as applied to physical readiness screens used in recruitment and military training.

Recommendation 6: The Army should evaluate Soldier performance for the future mission effectiveness of the TSU in the near term by leveraging existing research and development and by considering all DOTMLPF domains.

Mid to far term actions toward maintaining decisive levels of Soldier performance in TSUs include:

- Provide near-real time physiological readiness state reporting from Soldier and TSU to the command chain using physiological state monitors
- Leverage personality inventories, such as the Tailored Adaptive Personality Assessment System, to determine the cognitive, non-cognitive, and physical performance attributes that predict TSU performance.
- Conduct analyses to predict probable increased TSU measures of performance and measures of effectiveness levels attainable if two-year and five-year technology goals are met and anticipated improvements are implemented.
- Explore the potential to discern the state of the social network, morale, and other performance-relevant attributes from the communications among the TSU members without invading individual privacy and without individual identifications.

Recommendation 7: To maintain the currency of representative measures for the primary dimensions of Soldier and TSU mission performance, the Army, including its doctrine and training, research and development, acquisition and testing elements, should undertake a recurring program (at least biannual) to re-evaluate Soldier performance considering the analytical foundation for the functional design of the TSU, including numbers of Soldiers, grades and specialties, career experience, organization, and external support requirements.

Focusing on TSU Training

Focused training is essential to improving the performance of Soldiers and TSUs to levels that can assure overmatch. Not only will Soldiers and TSUs be expected to do more, but an increased emphasis on exploiting human-dimension knowledge demands innovative approaches. With the TSU as the centerpiece of future Army operations, small-unit leader training will be more important than ever.

Recommendation 8: The Army should focus training for the individual Soldier and TSU in the near-term as follows:

- Define TSU training objectives to produce TSUs that perform acceptably on the TSU measures of performance (MOPs) and measures of effectiveness (MOEs).
- Produce nonintrusive physiological status monitors to allow self-awareness and command chain assessments.
- Apply results of research in individual differences to the administration of TSU training.
- Expand sociocultural training capabilities to produce necessary TSU skills within time and resource constraints expected for TSU deployments.
- Expand instructor development to incorporate current theories of learning and feedback.
- Develop a concept for TSU master trainers to be assigned to company or battalion level to ensure continuous effective training of TSUs.
- Develop tools for TSU leaders (and leaders at higher levels) to assess Soldier and TSU training readiness against the TSU MOPs and MOEs.
- Ensure that effects of nutrition, hydration, sleep, dietary supplements, tobacco, and alcohol on cognitive and physical performance are incorporated in all modes of training of Soldiers and non-commissioned officers, including electronic games as well as live, virtual, and constructive simulations for individual (self) and group training.

Recommendation 9: In the mid to far terms, the Army should refine its focus on training for the individual Soldier and TSU by increasing the resolution of its suite of assessment tools to allow tracking of Soldier and TSU skill acquisition through and during each individual and collective training event, including live, virtual, and constructive simulations and electronic games.

Integrating the TSU into Army Networks

The Army has already recognized the important role of the network in achieving expanded capabilities in combat. Yet, dismounted Soldiers and TSUs today have limited organic capability (e.g., radios) to take advantage of networking in all mission environments. Ensuring full integration of the TSU into the Army network is essential to achieve decisive overmatch for dismounted TSUs and Soldiers.

A crucial concept to guide this integration is the necessity of ensuring that TSU leaders and individual Soldiers have sufficient *Situational Understanding*. Full situational understanding requires all three levels of situational awareness:

- Level 1 situational awareness is the perception of disaggregate elements of information acquired from data received from sensors either directly or indirectly; plus
- Level 2 situational awareness, often referred to as situational understanding, is achieved when Level 1 perceptions are further combined, interpreted, stored or retained for use by a Soldier or TSU, plus
- Level 3 situational awareness is reached when Level 2 perceptions are applied to project possible future events and anticipate outcomes.

Dismounted TSU leaders and Soldiers require send-and-receive access to communications networks, information networks, and sociocognitive networks. For decisive overmatch, the three types of networks must provide full use of sensor, lethal and other capabilities, both external and organic to the TSU,

Integration of the Soldier and TSU into the Army's networks will require near-term investments in Army networks, such as the following:

- Communications networks enhancements including TSU-level network management, remote control of radio transmission modes, and hands-free display interfaces capable of operating in all weather conditions, day and night, without compromising the security of the Soldier or TSU.
- Information networks capable of providing position location and tracking information in GPS-denied environments, automated tagging of information received to aid visualization, prioritization and dissemination, and access to level 1 situational awareness data from supporting sensors.
- Socio-cognitive networks capable of providing real-time access to such things as reports on tactical ground activities from collateral units and biometric databases for identification of adversaries.

Network capabilities required in the mid to far terms include the following:

- Integration with autonomous systems networks and user interfaces in addition to audible or digital interfaces, such as gesture recognition.
- Network applications, such as an intelligent TSU leader-assist tool to provide critical information alerts, assistance with planning and execution of missions, and automatic reporting and to provide behavior trend analyses of changes in enemy and civilian activities.
- Network enabled support of information sharing with collateral forces.

Recommendation 10: To achieve decisive overmatch capabilities, the Army should fully integrate the Soldier and TSU into existing and planned communications, information, and socio-cognitive networks ensuring that the network enhancements required for this purpose address all DOTMLPF domains.

Measures for assessing levels of situational understanding (MOPs and MOEs) would have utility for materiel development and evaluation, analytical modeling and simulation, human factors research, as well as TSU training. It is possible that physiological correlates to such measures could be confirmed, and limited instrumentation could be operational, for validation of materiel development trials conducted in the mid term. By the far term, it should be possible to assess the range, resolution, and reliability of Soldier and TSU situational understanding in relevant operational environments in real time.

Recommendation 11: In an immediate initiative, the Army should engage the S&T community (from both human and materiel perspectives), users, trainers, and other stakeholders in Army networks, to produce measures for assessing levels of situational understanding needed by the TSU.

Balancing TSU Maneuverability, Military Effects, and Survivability

In the context of what the Army expects a dismounted TSU to do—across all the missions and tasks anticipated in future unified land operations—overmatch requires a mission-appropriate balance of maneuverability, survivability, and military effects (including lethal, nonlethal, stability, and humanitarian effects). For dismounted operations, the fulcrum on which maneuver, survival, and effective action must be balanced is the Soldier's combat load. When the balancing act fails, the consequences degrade TSU and Soldier capability in all three areas. Based on presentations and discussions with Soldiers, the committee found that, in practice, the dismounted Soldier's combat load is far too great, often exceeding the upper limits stated in Army doctrine.

Excessive Soldier loads degrade not only maneuverability of both individual Soldiers and TSUs but also their resilience, survivability, and effectiveness. With such heavy burdens, traversing rough terrain and making rapid changes in direction, speed, and orientation greatly increase Soldiers' susceptibility to injuries. The load is excessive because the various subsystems and components of the Soldier and TSU systems are being optimized independently of each other.

Just as important for decisive overmatch are the potential benefits of getting the balance right. The Committee identified potential benefits for improved Situational Understanding; advantages in gaining and maintaining surprise or in immediately seizing the initiative even when an opponent acts first, through the ability to outmaneuver the opponent; more effective options for use of robot systems to support dismounted units; finding the right balance of body armor (individual protective equipment, or IPE) with other factors that contribute to Soldier load; and other benefits.

Materiel developers explained that IPE development and manufacturing programs go to great lengths to ensure sufficient sizes are available to effectively fit the diversity of body shapes and sizes in the Soldier population. However, they noted their surveys showed that a significant portion of solders in the field have been issued the wrong size, usually degrading their mobility to a very measurable degree. Improvement in assuring that sufficient fitting skill and knowledge as well as size inventory is required. A broad survey of TSU Soldiers will be needed to determine the appropriateness of fit of issued IPE. From those results, a development and engineering program may be warranted to improve the tools for fit determination at issue points (near term).

Experimental trials are needed to develop models for predicting the vulnerability of dismounted individual Soldiers and TSUs to engagement as a function of Soldier load and measures/indicators of individual/TSU mobility and agility such as dash speed (e.g., cover to cover). Engagement factors included in these trials should include visual detection, identification, and targeting of the opposing element in relevant combat-encounter scenarios (e.g., Blue Force-initiated contact, ambush of Blue TSU, urban/village setting with sudden transition from stability operation to kinetic fight). Environmental factors including terrain, elevation, and weather would be later parameters to add to the models and scenarios incorporated in the trials.

The types of engagements included in these trials need to cover the range of engagement scenarios that dismounted units may encounter in future unified land operations, including stability operations as well as conventional combat (offensive and defensive tasks). The goal should be to develop realistic, validated models for use in evaluating a wide range of current approaches and innovative concepts for managing Soldier load to achieve an optimal balance of TSU and Soldier maneuverability, military effects, and survivability.

Recommendation 12: The Army should initiate and maintain a program of experimental trials to inform improved models for assessing the effectiveness of dismounted Soldiers and TSUs as a function of Soldier load and measures/indicators of mobility and agility. The program should include an iterative process to explore innovative concepts for balancing TSU maneuverability, military effects, and survivability, as well as continuing exploration of more traditional approaches such as lightening individual items carried and offloading Soldier load onto robotic carriers.

Flexibility with respect to effective action becomes even more demanding when TSU mission objectives require a dismounted unit to be prepared to shift rapidly among traditional lethal combat, nonlethal means of projecting force, and stability objectives where effectiveness is measured in terms of communication with the local population, building capacity for civil operations, or humanitarian objectives. Little is known about the effects of nonlethal weapons on adversaries or about their impact on engagement decision complexity for the Soldier. The effectiveness of nonlethal actions used as an alternative to lethal means will depend to a great extent on the perceptions of those being confronted.

Recommendation 13: In the mid-term, the Army should undertake research to identify a range of unambiguous signals of nonlethal intent. The research should extend to the exploration of cultural differences in intent interpretation.

Given the range of missions and tasks that dismounted TSUs may be called upon to perform in the future, even experienced leaders at the TSU level and higher echelons cannot be expected to know immediately the best combination of available options, extending across all DOTMLPF domains, for the optimal balance of maneuverability, military effects, and survivability in every environment and engagement. An easy-to-use mission planning aid could incorporate the relationships among options learned from prior operational experience (lessons learned), as well as the relationships among metrics, indicators, and DOTMLPF options found and validated through experimental trials and incorporated in assessment models used by the development community.

Properly designed, such a mission planning aid would include long distance endurance and sprint speed (as factors in engagement vulnerability), functions of terrain, meteorological factors, ration intake, loads, physical attributes of TSU members, and resupply points. It would identify the TSU member-by-load combination most likely to be the mobility limit for the TSU formation. If the empirical basis could be developed, the planning aid could also predict the probability that the mission would contribute to the long-term injury or disability of particular TSU members.

The mission planning aid would be used in training TSU leaders on the factors that affect squad mobility, including terrain, meteorological conditions, loads, load configurations, accumulated fatigue, individual protective equipment (IPE), and how factors like IPE fit and load configuration constrain agility. Practical exercises for leader trainees would increase confidence in using the planning aids in operations. Also, the aids to Soldier load planning and mobility and endurance effects of different loads could be incorporated in training simulations and games.

TSU leaders and their commanders at higher echelons need to understand how factors across all the DOTMLPF domains affect not only Soldier load but also the more encompassing goal of balancing maneuverability, effective action, and survivability to ensure small units have decisive overmatch wherever and under whatever circumstances they operate.

Recommendation 14: The Army should develop a mission planning aid to assist in balancing maneuverability, military effects, and survivability, for use in training and operations by TSU leaders and leaders at higher echelons.

Leveraging Advances in Portable Power

As long as electronics are used to enable the key equipments on which Soldiers and TSUs depend, advances in portable power will continue to provide decisive-edge potential to U.S. Soldiers. Power issues have doctrinal implications because of their impact on TSU tactics, techniques, and procedures. The last decade has seen major advances in portable power materiel technologies, which could have outsize influence on overmatch. However, this can occur only if the Army can leverage the advances to their full effect, which requires considering the full range of DOTMLPF implications for alternative portable-power solutions.

While the Army is well on the way toward implementing a rechargeable battery technology that could become the primary energy source for the Soldier on the battlefield, aside from the materiel development itself, critical DOTMLPF elements have not been evaluated. There is no doctrinal philosophy for the TSU to recharge the battery; there is no organizational equipment to support recharging; there is no hint of the training required; there is no parallel materiel development of a recharger or fuel reformer to exploit new rechargeable battery or fuel-cell technologies.

Rechargeable lithium-air energy sources used as the primary energy source in hybrid configurations have high potential to replace many of the disposable and rechargeable storage systems now in use. The selection of rechargeable battery storage technology as the primary choice for the Soldier energy source necessitates the parallel introduction of a recharger technology sufficiently small and lightweight that would be applicable at the squad level. Successful development of a JP-fuel reforming technology would allow for small combustion engine battery chargers of low cost and light weight. The Army needs to complete development of JP-reforming technology over a wide range of sizes in order to exploit either rechargeable battery technology or fuel-cell technology.

Advances in portable power will contribute to the decisiveness of TSUs by giving future Soldiers high confidence that their equipment ensemble will have sufficient energy to carry out the mission. Achieving this goal will help to reduce fatigue, eliminate the anxiety associated with resupply, increase confidence in situational awareness from powered sensors, and assure communications links with higher levels in the command structure.

Recommendation 15: The Army should develop and maintain a robust program in advanced energy sources based on full analysis of DOTMLPF elements with the goal of eliminating power and energy as limiting factors in tactical small unit operations.

1

Introduction

This chapter provides background and context for the Statement of Task (SOT) and the committee's study approach. It discusses important study assumptions and/or limitations, the approach taken to achieve the study objectives, and the organization of the report.

ORIGIN OF THE STUDY

In March, 2010, the leadership of the National Research Council (NRC) Board on Army Science and Technology (BAST) met with the Assistant Secretary of the Army for Acquisition, Logistics, and Technology [ASA(ALT)] to discuss how to make the Soldier[1] more decisive on future battlefields. At the request of the Assistant Secretary, Dr. Malcolm O'Neill, the next three BAST meetings included presentations related to the individual Soldier and small unit operations. As a result of these board meetings, the BAST recommended that an unclassified study was needed to review and examine technology areas with potential to make U.S. Soldiers in small units decisive and dominant on future battlefields.

In accordance with the BAST recommendation, the ASA(ALT) approved the SOT shown in Box 1-1 and requested that the NRC establish an ad hoc study committee consisting of experts in appropriate fields to accomplish the study tasks under the oversight of the BAST.

Areas of Focus

This section defines terms used in the SOT, describes assumptions made by the committee, and clarifies the areas of focus for the study as agreed upon with the ASA(ALT) sponsor. At the beginning of the study, the committee chair

[1]In December 2003, Army Chief of Staff GEN Peter J. Schoomaker directed that all command information products, including base newspapers, capitalize the word "soldier" to give U.S. Army Soldiers "the respect and importance they've always deserved" (Coon, 2003). GEN Schoomaker's directive has become standard practice in Army communications, including doctrine documents. Although it is not the general policy of the National Academy Press or the editorial staff of the NRC, the committee requested and received permission to capitalize "Soldier" in this report when referring to U.S. Army soldiers.

BOX 1-1 Statement of Task

The U.S. military does not believe its soldiers, sailors, airmen, and marines should be engaged in combat with adversaries on a "level playing field". Our combat individuals enter engagements to win. To that end, this country has used its technical prowess and industrial capability to develop decisive weapons, weapons that over-match those of potential enemies, such as the M1A2 tank, the F-22 fighter, and the Seawolf attack submarine. The country is now engaged in what has been identified as an "era of persistent conflict" in which the most important weapon is the dismounted soldier operating in small units. More than for soldiers in Vietnam, Korea, and WWII, today's soldier must be prepared to contend with both regular and irregular adversaries. Results in Iraq and Afghanistan show that while the US soldier is a formidable fighter, his contemporary suite of equipment and support does not enjoy the same high degree of overmatch capability exhibited by large weapons platforms—yet it is the soldier who ultimately will play the decisive role in restoring stability.

A study is needed to establish the technical requirements for overmatch capability for dismounted soldiers operating individually or in small units. What technological and organizational capabilities are needed to make the dismounted soldier a decisive weapon? How can technology help those soldiers remain decisive on a changing, uncertain and complex future environment? The study will examine the applicability of systems engineering to soldiers and small units, as well as specific technology areas that are relevant to making soldiers decisive, particularly in conditions where we still take casualties today (movement to contact and chance encounters). Technology areas to be considered should include (but not be limited to) situational awareness, weapons, mobility, and protection, adaptation to battlefield environments (e.g., clothing, cooling), communications and networking, human dynamics (e.g., physical, cognitive, behavioral), and logistical support (e.g., medical aid, food, water, energy).

The NRC will establish an ad hoc study committee to examine these requirements. The committee will:

1. Determine the elements of overmatch capabilities necessary for a dismounted soldier to be a decisive weapon on the battlefield, Consider both the individual soldier as well as the soldier as part of a small (squad-size or smaller) unit.
2. Identify technical requirements for optimizing soldiers and small units to achieve overmatch capabilities on the battlefield. Consider technology and societal trends that may affect the balance between U.S. forces & adversaries both now and in future years.
3. Identify near-term, mid-term and far-term technologies in which new or enhanced S&T investments would facilitate the development of decisive soldier capabilities.
4. Determine the relative importance of such investments in making the soldier decisive on future battlefields.

met with the Army Chief Scientist to clarify the scope of the study. The following guidance resulted from that meeting.[2]

> The study will address the operations by dismounted infantry Soldiers and squad-size or smaller units in the future and include the full

[2]Quoted text is from the discussion paper used in the meeting between Dr. Scott Fish, Army Chief Scientist, and LTG (U.S. Army, retired) Henry J. Hatch, Chair, Committee on Making the Soldier Decisive on Future Battlefields, April 5, 2011.

> spectrum of operations[3] extending from stable peace to general war in an environment of persistent conflict. The latter introduces the requirement that the Soldier be resilient—physically and mentally. 'Future land operations' exclude the current operations in Iraq and Afghanistan although the study will draw on those and other experiences as appropriate.
>
> The primary focus of the study will be the equipped, trained, and supported (full Doctrine, Organization, Training, Materiel, Leadership and Education, Personnel and Facilities) individual Soldier, and the study will address both materiel and human capabilities. The human component includes physical (health and fitness), mental, and social attributes including cognitive (knowledge and skills) and noncognitive (temperament, strength of character, social awareness, etc.). Each Soldier is unique, and talents and traits, including cognitive abilities, will vary considerably among the Soldiers in a particular unit. The Soldier is an integrated system—materiel and human.
>
> The intent is to avoid the traditional approach of "building material systems around the Soldier," and to facilitate the development of an Army composed of Soldiers and small units that can be adapted to whatever mission is assigned.

This clarifying guidance became crucial to the study approach adopted by the committee, serving as a touchstone for affirming what many readers of this report may initially see as controversial premises assumed and positions defended.

Definitions

The SOT contains several terms that are defined below for the purposes of the study:

- Decisive: An adjective that refers to the ability to settle or decide an outcome; to be conclusive. The focus for this study is on making Soldiers in small units the successful decisive element on future battlefields and other areas of operations. Soldiers must take decisive action in the continuous, simultaneous combinations of offensive, defensive, and stability or defense support of civil authorities tasks (U.S. Army, 2012, Pp. 5-6). In doing so, Soldiers and the tactical small unit (TSU) must be effectively inside the opponent's decision-making cycle to act preemptively and not just reactively. For Soldiers and TSUs to be decisive, they must overmatch their opponents in all missions.

---

[3]"Full spectrum operations" was the Army's operational concept at the time the SOT and the clarifying guidance were drafted (U.S. Army, 2011a). As of November 2011, "unified land operations" became the Army's operational concept, and "range of military operations" replaced "spectrum of conflict" (U.S. Army, 2011b; U.S. Army, 2012). See Appendix C for current doctrinal terminology.

- Overmatch: To be more than a match for an opponent by greatly exceeding comparative measures of the opponent's capabilities. Overmatch implies sufficient superiority to ensure operational success.
- Near, mid, long terms: For this report, the near term is within 5 years, mid-term is 5-10 years, and long term is beyond 10 years.

STUDY APPROACH

The SOT charged the committee with identifying both technological and organizational capabilities that are needed to give a dismounted small unit decisive overmatch against future adversaries. Technical requirements for optimizing Soldiers and small units to achieve overmatch were to be identified in areas including but not limited to situational awareness, communications and networking, weapons, mobility, protection, human dynamics, and logistical support. To address its charge, the committee conducted an extensive information-gathering effort that included visits to Army training facilities and service laboratories, presentations from and discussions with Army leaders from both the requirements development and acquisition communities, and meetings and interviews with Soldiers and small unit leaders recently returned from deployment in overseas operations. The committee also identified and reviewed key publications in the open literature and shared the individual members' experiences and expertise with representatives of both the Army science and technology community and the operational Army. The committee's formal meetings and site visits are listed in Appendix B.

From the outset of the study, the committee noted that the soldiers and tactical small units in Iraq and Afghanistan were expected to perform in a variety of operations in addition to traditional combat and that the "battlefield" had become far more complex. These additional roles, referred to as wide area security and combined arms maneuver, were discussed with the Army during data-gathering and later articulated in documents that were released during the study. (U.S. Army, 2011b; U.S. Army, 2012).

In the revised doctrine, the range of military operations holds infantry soldiers and tactical small units equally as responsible for stability operations, such as wide area security, as they have been in the past for offensive and defensive combat operations alone. The increased scope of responsibilities provided the committee with perspective on what would be needed to achieve decisive overmatch in the future, and it also affected the approach to the study.

As a result, it was clear to the committee that the Army had begun a transition to the future TSU in its current operations and that the TSU would depend much more heavily on the abilities of Soldiers in the future. The study approach would have to evaluate existing and contemplated technologies in light of this expanded operational mission set for the dismounted Soldier and TSU.

Essential Principles to Achieve TSU Overmatch

During the course of the committee's deliberations during its information-gathering activities and in the subsequent report-drafting phase, four principles kept emerging that, to the committee, seemed essential for the Army to embrace if it is serious about providing dismounted Soldiers and small units with decisive overmatch across the range of potential missions and tasks envisioned for future unified land operations. Without rigorous adherence to these principles, the "optimization" of Soldiers and small units to achieve overmatch across the expanded range of military operations, will not be achieved, no matter what piece-part technologies are developed to facilitate "decisive Soldier capabilities."

1. The *human dimension*, as defined by the Army, needs to be expanded in scope, and more emphasis needs to be placed on this expanded concept of the human dimension and other non-materiel aspects of potential solutions to provide overmatch capability. The committee's view of what the human dimension should include is discussed below.
2. The complexity of what the dismounted Soldier does and of the means available to accomplish those tasks requires that the Soldier be viewed as a system in which components and subsystems must work together seamlessly and without interference with or diminishment of other functions of this Soldier-system. The committee thus agrees with the assertion, made by the Army and advanced in numerous prior reports to the Army, that the Soldier is a system—albeit a *human-based* system unlike *platform-based* systems such as tanks, submarines, or fighter aircraft. If the Soldier is a human-based system, then a dismounted TSU is a system of these Soldier-systems. From this perspective, a comprehensive, analytically based systems engineering capability is essential to evaluate and make trades among capability options that encompass all the domains of TSU performance captured under the military rubric of DOTMLPF: Doctrine, Organization, Training, Materiel, Leadership and Education, Personnel and Facilities.
3. Metrics (e.g., attributes, measures of performance, and measures of effectiveness) for the Soldier and TSU are needed at the outset to guide and measure actions to provide capability improvements in all facets of the Soldier and TSU acquisition life cycles (conceptualization, development, test and evaluation, training, etc.)
4. The Army's acquisition system needs to be integrated, streamlined, and tailored to embrace the three principles above and to ensure that solutions identified through a systems engineering methodology are developed, tested, and delivered in an expeditious and efficient manner.

The committee devoted considerable time and energy to distilling and illustrating these four principles while continuing its efforts, in response to the

SOT, to identify specific technologies that seemed to have the greatest potential to contribute to the ultimate overmatch solution. Although the SOT did not explicitly charge the committee with exploring these overarching principles, the committee sees them as an essential part of any adequate answer to the stated question, "How can technology help those Soldiers remain decisive on a changing, uncertain, and complex future environment?"

Committee's Approach to the Human Dimension

The first essential principle enumerated above calls for an expansion of the Army's current conception of the human dimension. As it came to understand the non-materiel side of TSU and Soldier capabilities, the committee decided that the greatest returns on Army investments for improvements in the near, mid, and far terms would be achieved by integrating the materiel aspects of technology developments with non-materiel aspects found primarily in the human dimension. The committee learned during its study that there are known advances in individual and collective human performance that offer potential to meet the identified capability needs of future dismounted operations but that have not been applied by the Army.

Decisive overmatch capabilities will only be achieved if adequate attention to and investment in human dimension solutions are fully coordinated with solutions from the materiel dimension. The dimensions cannot be applied in isolation; the dismounted TSU and the Soldier will only have decisive overmatch when both dimensions come together in capabilities superior to those opposing them, across the full range of missions and tasks expected in unified land operations.

This assessment is consistent with the opening 50 pages of the seminal study report on the human dimension by the U.S. Army Training and Doctrine Command (TRADOC), where a half dozen quotations put Soldier attributes ahead of weapons in determining battle outcome (TRADOC, 2008a). From Abrams and Ardant du Picq through Marshall and Patton to Sun Tzu and Van Riper, the best military minds have expressed in their own way the view stated in the Army's 2005 edition of its capstone document, *The Army* (Field Manual 1):

> First and foremost, the Army is Soldiers. No matter how much the tools of warfare improve; it is Soldiers who use them to accomplish their mission. Soldiers committed to selfless service to the Nation are the centerpiece of Army organizations.
>
> (U.S. Army, 2005, p. 1-1)

That TRADOC study report, and the more summarizing TRADOC concept document that followed (TRADOC, 2008b), used a definition of "human dimension" more suited to a lay audience or the popular press than to a scientific study or disciplined analysis: "... the human dimension encompasses the moral, physical, and cognitive components of Soldier, leader, and organizational development and performance essential to raise, prepare, and employ the Army in full spectrum operations" (TRADOC, 2008a, p. 9). Throughout that study, and especially in the chapter devoted to the moral component, the meaning of "moral" is extended beyond its dictionary definition to include unit morale and cohesion, language skills, and cultural awareness (TRADOC, 2008a). The chapter on the cognitive component of the human dimension is devoted exclusively to training. Neither selection of recruits nor placement of Soldiers with different cognitive attributes is discussed. Finally, Soldier attributes of temperament or personality were not discussed in either TRADOC document, but these attributes have a profound influence on the success of the ubiquitous "strategic corporal" and the collective squad.

In presentations from Army briefers, the committee frequently heard the term "human dimension" or read it on briefing slides, but the usage implied little beyond "other-than-materiel." A more useful characterization of what should be included in the human dimension occurred in a two-page information paper on the human dimension written in June 2011 and approved by the Chief of the Human Dimension Task Force (Johnson, 2011). The paper refers to "cognitive, physical, and social" attributes of Soldiers; accession and selection of personnel; training and education; Soldier readiness; development of Soldiers, leaders, and organizations; and individual and unit resiliency (Johnson, 2011). The range of topics and their implied interdependency in this brief statement are closer to the committee's working concept of the human dimension.

The committee recognized that high standards of morals and ethics are essential if Soldiers are to successfully prosecute all wartime missions, but it assumed that topics relating specifically to moral conduct and ethical performance of Soldiers were outside the scope of its task statement.

Another useful source for what should, in the committee's view, be included in the province of the human dimension is a three-page article by a retired Army major general on "The Human Dimension in the Close Fight" (Scales, 2012). The author asks the reader to imagine being among the Soldiers of a dismounted squad on night patrol when they are suddenly ambushed just beyond range of supporting fires from their forward operating base. The human attributes that this author views as making "soldiers and leaders into superbly competent small units" include seeing and sensing the enemy so there are no surprises, emotional stability under the stress of combat, group resilience, the ability of leaders to make sound decisions quickly under extreme stress, and trust in support from other units and higher echelons. The author concludes with the following recommendation for winning the close-combat fight in the future:

> ...I believe now more than ever that the best investment we can make of our diminishing human and capital resources would be to use the human sciences to improve the fighting power of close-combat soldiers, to focus as much on what goes *in* the soldier as what goes *on* the soldier.
>
> (Scales, 2012, p. 38)

In light of both the official TRADOC sources cited first and the unofficial, informal, but more comprehensive accounts of the human dimension suggested in the latter two cited references, the committee has used the following as its working definition, for purposes of this study:

> As used in this study, the human dimension means all of the attributes of the individual Soldier and of the collected Soldiers of the TSU that impact performance of mission tasks. These include the skills, abilities, and knowledge brought with them into the Army upon recruitment, even from prior education or job experiences; personality traits; individual and collective military training; skills, abilities, and knowledge from prior military assignments; TSU command chain leadership; unit social environment including morale, cohesion, and emotional state; the ergonomic design or human factors engineering of the Soldier-machine interfaces; as well as locale acclamation (time zone, elevation, temperature, etc). Skills, abilities, and knowledge include the physical, mental, and emotional. Bearing with real impact but less directly on mission task performance are the domestic or family environments of each Soldier, which have not been included here. Nor has the committee included issues of morality that may bear on overall mission accomplishment in a strategic sense, but not on tactical tasks, except as morality issues may influence the effectiveness of the unit leadership chain or the health of the unit social environment.

When this working definition of the human dimension is applied to the dismounted Soldier and dismounted TSU as human-based systems that require a systems engineering approach, the result is similar in purpose to that of concepts such as "human-systems integration," as presented, for example, by Booher (2003). However, as discussed further in Chapter 3, military implementations to date of human-centered systems design, such as the Army's MANPRINT program or the Air Force's Human Systems Integration Office, have fundamental constraints and flaws that limit their relevance as models for the approach advocated in this report.

REPORT ORGANIZATION

Chapter 1 (Introduction) provides the background of the study, the committee's approach to addressing its SOT, an initial discussion of key concepts, and the report organization. Chapter 2 (Capabilities) uses the committee's view of the missions and tasks that are likely to be assigned to dismounted TSUs and Soldiers in future operations to identify critical capability needs and opportunities to achieve decisive overmatch. Chapter 3 (Setting the Conditions to Achieve Soldier and TSU Overmatch) discusses the essential conditions necessary for designing, developing, and implementing technologies that will ensure the dismounted TSU is decisive on future battlefields. Chapter 4 (Achieving

Overmatch) applies the approach recommended in Chapter 3 to discuss prospective near-, mid-, and long-term options for inclusion in the systematic analysis process. In particular, it focuses on five areas for improving TSU capability that the committee agreed have the greatest potential for contributing to decisive overmatch.

REFERENCES

Booher, H.T. 2003. Handbook of Human Systems Integration, New York, N.Y.: John Wiley and Sons, Inc.

Coon, C. 2003. Soldier—and that's with a Capital 'S'. Stars and Stripes. Available online http://www.military.com/NewContent/0,13190,122303_Soldier,00.html. Accessed March 20, 2013.

Johnson, T. 2011. The Human Dimension. Approved by COL Steven Chandler, Chief, Human Dimension Task Force.

Scales, R.H. 2012. The human dimension in the close fight. Available online www.ausa.org/publications/armymagazine/archive/2012/05/Documents/Scales_0512.pdf. Accessed September 27, 2012.

TRADOC (U.S. Army Training and Doctrine Command). 2008a. TRADOC Pamphlet 525-3-7-01: The U.S. Army Study of the Human Dimension in the Future: 2015-2024. Arlington, Va.: Headquarters, U.S. Army. Available online www.tradoc.army.mil/tpubs/pams/p525-3-7-01.pdf. Accessed September 27, 2012.

TRADOC. 2008b. TRADOC Pamphlet 525-3-7: The US Army Concept for the Human Dimension in Full Spectrum Operations: 2015-2024. Arlington, Va.: Headquarters, Department of the U.S. Army. Available online www.tradoc.army.mil/tpubs/pams/p525-3-7.pdf. Accessed September 27, 2012.

U.S. Army. 2005. Field Manual 1 (FM 1). June 2005. Arlington, Va.: Headquarters, Department of the Army. Available online www.army.mil/fm1/. Accessed September 26, 2012.

U.S. Army. 2011a. Operations. Army Field Manual (FM) 3-0, Change 1. Arlington, Va.: Headquarters, Department of the Army.

U.S. Army. 2011b. Army Doctrine Publication (ADP) 3-0. Unified Land Operations. Arlington, Va.: Headquarters, Department of the Army. Available online http://armypubs.army.mil/doctrine/ADP_1.html. Accessed March 22, 2013.

U.S. Army. 2012. Army Doctrine Reference Publication 3-0 (ADRP 3-0). Unified Land Operations. Available online http://armypubs.army.mil/doctrine/ADRP_1.html. Accessed March 20, 2013.

2

Capabilities

This chapter discusses desired capabilities for the dismounted tactical small unit (TSU) and the Soldiers in these TSUs, organized by major areas of TSU function. It describes the most critical capability needs identified by the committee for future dismounted TSUs and Soldiers to have decisive overmatch (as explained in Chapter 1) across the range of military operations.

In the absence of stated formal requirements for the dismounted TSU, from which capabilities could be derived, the committee gathered its information directly from briefings presented by members of the Army staff and other information-gathering activities (see Appendix B), from Army publications and weblinks, and from unclassified public sources.[1] Particularly important were the discussions with Soldiers, both enlisted and officers, who had recently returned from duty in small infantry units in military operations. Even without a formal requirement stating TSU capabilities, it was apparent to the committee that the TSU is much more than a "formation" of individual Soldiers and that there is no documented analytical foundation on which to base TSU capability needs. Lacking the benefits of such analysis, the committee began its study by reviewing the missions and capabilities of the current dismounted Army squad and identifying areas where existing squad performance could be improved, whether the improvements were in materiel, in some aspect of the human dimension as defined in chapter 1, or (more frequently) through solutions that combined improvements from both the human and materiel dimensions.

For the purpose of organizing TSU and Soldier capabilities, the committee used the following five categories of general operational capability. The committee believes that removing deficiencies and taking advantage of capability opportunities in these five areas is the key to giving future dismounted TSUs and the Soldier decisive overmatch across the range of missions and tasks the Army has envisioned for them.

- **Situational Understanding** includes (a) the aspects of sensory/perceptual data reception usually associated with situational awareness (referred to

---

[1]TRADOC representatives told the committee at its initial meetings that the Army was preparing an Initial Capabilities Document (ICD) for the tactical small unit that the committee could use as a foundation for TSU requirements in the study. The ICD effort was apparently discontinued.

below as Level 1 situational awareness), plus (b) the situational understanding achieved when a TSU or individual combines, interprets, stores, and retains Level 1 data (this is referred to as Level 2 situational awareness below), plus (c) the use of that understanding to project possible future events and to anticipate their outcomes (Level 3 situational awareness).

- **Military Effects** include any of the effects needed to accomplish the offensive, defensive, and stability tasks that may make up a dismounted operation for a TSU. Among the military effects for which the TSU must be prepared are the following:
  - ***Lethal Effects***. Physical destruction of equipment or infrastructure, killing or wounding of personnel in an opposing force, and other damaging actions that create permanent or near-permanent damage
  - ***Nonlethal Effects***. Temporary incapacitation of equipment, infrastructure, or personnel, while minimizing fatalities, permanent injury to personnel, and undesirable damage to equipment, property, and the environment
  - ***Stability (Population) Effects***. Effects relevant to the objective of "winning the hearts and minds" of the people living in the theater of operations[2]
  - ***Stability (Capacity-Building) Effects***. At the TSU level (the "village" or "city sector" level), capacity-building effects[3] include changes in local (village, city sector, etc.) security, governance, rule of law, and socioeconomic capacity. Positive changes are the types of effects sought through "Village Support Operations," which aid in defeating irregular threats imbedded in noncombatant populaces
  - ***Humanitarian Effects.*** Improvements in health, medical, nutritional, and living conditions
- **Maneuverability** includes both agility and mobility, with a focus primarily on physical capability. Mobility includes the ability to move from point to point across/through various types of natural and manmade terrain, including traversing obstacles, in all weather and light conditions. Agility includes the ability to quickly and significantly change one's direction, speed, body orientation, and weapon orientation. (Note: mental agility—the ability to think and draw conclusions quickly, intellectual acuity—is also extremely important to the TSU and the Soldier. Mental

---

[2]Current Army doctrine defines four *stability mechanisms* for affecting civilians in order to attain conditions that support establishing a lasting, stable peace: compel, control (impose civil order), influence, and support (U.S. Army, 2012, p. 2-10). Of these, the "influence" and "support" mechanisms are the two most likely to utilize stability (population) effects.

[3]A general description of *capacity building* not specific to dismounted TSU/ Soldier operations would be as follows: Capacity building is the building of human, institutional, and infrastructure capacity to help societies develop secure, stable, and sustainable economies, governments, and other institutions through mentoring, training, education, physical projects, the infusion of financial and other resources, and most important, inspiring people to improve their quality of life.

agility is not subsumed under Maneuverability; rather, it is a Soldier and TSU capability that is supported and extended by Level 3 situational awareness and in turn supports decision-making for military effects.)

- **Sustainability** pertains to the sustainment warfighting function, which the Army defines as "the related tasks and systems that provide support and services to ensure freedom of action, extend operational reach, and prolong endurance" (U.S. Army, 2011).
- **Survivability** includes both physical and mental survival. From a physical perspective, survivability includes not only protection from threat weapons (i.e., avoiding physical damage to Soldiers or the TSU) but also reducing the ability of the threat to detect, attack, or hit the Soldier and TSU. It also includes the ability to prevent death or permanent loss of bodily functions (including traumatic brain injury) if a Soldier is damaged. Each Soldier must also be protected from natural threats such as extremes in temperature, insects, and infectious agents. From a mental perspective, survivability means that each Soldier must have cognitive functions protected— i.e., maintain mental/psychological resilience (this includes avoiding/preventing post traumatic stress disorder).

Because these categories of capability are defined to be comprehensive, they overlap and complement each other. Consider, for instance, the role of decision-making in military effects, which depends on situational understanding. As described in two recent reports on science and technology that would contribute to stabilization and reconstruction operations, the adaptive and decision-making challenges for tactical leaders have grown tremendously with increased complexity of operational environments and range of military operations (Chait et al., 2006; 2007). These environments—particularly when an opposing force elects to pursue irregular warfare in the midst of U.S. Army stability operations— require TSU leaders who are superbly adept at utilizing a range of new operational procedures and technologies but are also keenly aware of and attuned to the entire social, political, economic, and cultural context in which their decisions are made and have effects. For decisive overmatch in these complex adaptive environments, how can the Army best prepare TSU leaders to function effectively? Are current leader selection, training, and development programs giving the Army agile and effective leaders for operations where irregular warfare is an ever-present threat, while dispersed, dismounted units are also pursuing stability and humanitarian effects as primary mission objectives?

TSU MISSIONS AND TASKS

The dismounted TSU—today's infantry squad—must be competent in both combined arms maneuver and wide area security. These core competencies are demonstrated through continuous, simultaneous combinations of offensive, defensive, and stability tasks (U.S. Army, 2012, p. 2-8; see Appendix C for relevant excerpts defining and illustrating these tasks).

> . . . .While all operations consist of simultaneous combined arms maneuver and wide area security in various proportions, most tactical tasks will be predominantly characterized by one or the other. The preponderant core competency determines the choice of defeat or stability mechanisms to describe how friendly forces accomplish the assigned mission. Generally, defeat mechanisms [employed in offensive and defensive tasks] are appropriate for combined arms maneuver, while stability mechanisms are best suited for wide area security.
>
> (U.S. Army, 2012, para. 2-33, p. 2-9)

Wide area security includes friendly force security, as well as security of the local population and infrastructure, support to law enforcement, support to reconstruction operations, and other stability-related activities.

To exercise both combined arms maneuver and wide area security with decisive superiority, the TSU needs capability for all of the following activities:

1. Deny anti-access and area denial capabilities of enemy forces, including criminal elements;
2. Conduct reconnaissance in close contact with civilian populations;
3. Rapidly transition from one operation to another within the same day and within a small geographical location—for example, support humanitarian, security, and combat operations simultaneously;
4. Conduct and sustain operations from and across extended distances and in austere environments;
5. Understand complex situations with the potential for both lethal and nonlethal engagement, especially with respect to how information operations and diplomatic, military, and economic activities may have consequences extending across political, military, economic, security, information, and infrastructure domains;
6. Understand the relationship of local operations to higher level operational and strategic goals and ascertain how best to achieve unity of effort in supporting these goals;
7. Understand the human terrain of its operating environment, which includes the diverse civilian population (national and foreign) and coalition partners with whom the Soldiers of the TSU will interact; and
8. Conduct sustained efforts to build partner capacity, prevent conflict, and prepare for contingencies.

Additionally, the TSU must accomplish its missions under various constraints such as rules of engagement, a decentralized enemy blending with the civilian population, the need to work with coalition and local security forces, and a distributed battlefield.

The role of small Army units (companies and below) in accomplishing the wide range of potential combat (offensive/defensive) and stability tasks has become more critical over time, as illustrated in Figure 2-1. Additionally, a small unit's area of operation has been increasing substantially as the context of warfare has shifted from the mechanized, state-on-state conflict of the first Persian Gulf War and earlier land wars (including the Cold War preparations for Soviet invasion in Central Europe) to non-state players using guerilla and terrorist tactics. As an example, around the year 2000, the area of operation of a brigade combat team (BCT) was approximately 2,700 square kilometers. In 2011, the 4th BCT, 10th Mountain Division, was responsible for 13,000 square kilometers (AUSA, 2011).

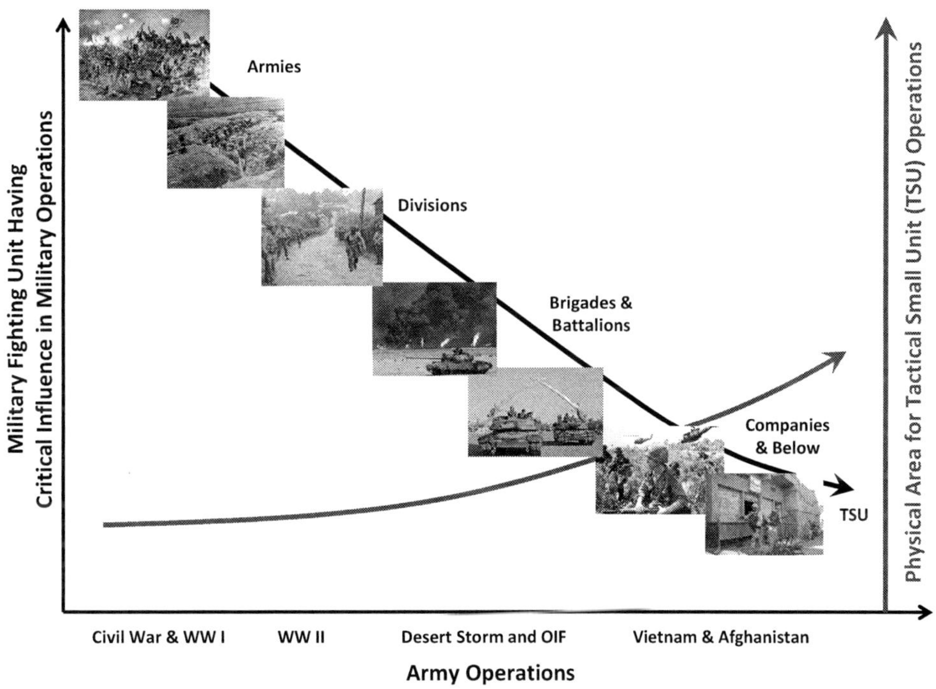

FIGURE 2-1. Decreasing size of fighting unit with critical influence and increasing area of operation for a tactical small unit. SOURCE: Based on AUSA (2011) and Dr. Marilyn Freeman, Deputy Assistant Secretary of the Army for Research and Technology, "Providing Technology Enabled Capabilities to Soldiers and Tactical Small Units," presentation at the 2011 AUSA ILW Winter Symposium and Exposition, Fort Lauderdale, Florida, February 23, 2011.

To make the TSU and Soldier decisive on the battlefield, capabilities are needed to enable TSUs and Soldiers not only to dominate opponents in lethal and nonlethal engagements but also to sustain other operations, including those with stability and humanitarian effects, for long periods of time before, during, and after such engagements. General capability enhancements are needed in areas of situational understanding, maneuverability, military effects, sustainability, and survivability as described below. These enhancements must be understood in the context that today's TSUs already possess a high level of capability in each of these general areas. A TSU can move almost anywhere, under almost any condition. When needed, current TSUs can provide a high volume of lethal small arms fire for a short time in any direction. This small arms fire can often (but not always, and even less often in in a timely manner!) be integrated with supporting direct and indirect fires. Similarly, with its nine or more pairs of human eyes actively and intelligently "sensing," a TSU walking through a village market today can gain a tremendous amount of information about the interactions among the local population, their reactions to the TSU, and their apparent well-being.

However, neither movement nor lethal fires nor simple presence is exclusively decisive. An integrated combination of fire and movement (in a close combat lethal engagement) or of community presence and noncombatant conversation (in a stability engagement) can often win that engagement, but much more is needed to achieve decisive TSU effectiveness for mission objectives. In addition to lethality and mobility, network integration, protection, socio-cognitive performance, and other materiel and human dimension components of capability together are needed to give a dismounted TSU decisive overmatch in all actions. The next five sections of this chapter explore the TSU and Soldier capabilities required in these missions and tasks by focusing on one capability area at a time: situational understanding, military effects, maneuverability, sustainability, and survivability. After that exploration of required capabilities, the committee discusses current capability weaknesses and opportunities, many of which cross over all or several of these capability areas, which the Army should address to ensure that future TSUs and Soldiers have decisive overmatch across the entire range of military operations.

The identification of specific capabilities for the future TSU depends on many unknowns relating to future Army operations. Therefore, to simplify its analysis and discussion of capability needs, the committee made the following assumptions:

- The TSU is not restricted to its current nine-man infantry squad organization and may be augmented by additional Soldiers or equipment in the future.
- The TSU will remain as a part of a platoon, which will remain as part of larger organizations.
- The TSU will be supported by assets at the platoon, company, battalion, and brigade-and-above levels.
- The TSU and individual Soldiers will need to sustain operations over long periods of time (e.g., 3 days as a dispersed and decentralized force, and 8 days as part of a larger force).

SITUATIONAL UNDERSTANDING

This section begins with the role of decision-making in achieving overmatch and how individual and shared situational understanding is essential to decision-making and execution. In that context, it then explains how three levels of situational awareness are needed to support decision-making and execution that produces decisive action.

The Role of Decision-Making in Overmatch

Deliberate decision-making is the process of identifying a problem to be solved, developing alternative courses of action for consideration, comparing anticipated outcomes of those courses of action, and selecting a course of action from that set for execution. It is critical to acknowledge that making decisions well is one, if not *the*, central goal for the dismounted Soldier and TSU with decisive overmatch. The challenge, of course, is that Soldiers and TSUs must make these decisions (1) under conditions of limited information, (2) when they have only limited time to make their decision, and (3) under conditions in which outcomes are uncertain (although it should be noted that as long as the likelihood of an outcome is known, this poses no special problem). Further, in order for the TSU to achieve decisive overmatch in a direct engagement, the timing of the decision must be within the opponent's decision cycle; that is, it must be made and acted upon more quickly than the time for the opponent to react.

In practice, Soldiers and TSUs achieve this kind of deliberate decision-making in one of two basic ways; they either engage in some kind of formal reasoning or they come to this decision with a more intuitive, or "gut-level" process. It is critical to acknowledge that both methods can work well. In fact, perhaps the central goal of training is to teach Soldiers and TSUs how to make decisions rapidly using intuitive approaches that are built by effective training rather than by slower, formal-reasoning-based techniques. Current neuroscientific research makes it clear that practice does indeed foster this change, shifting the location of decision-related brain activity from the frontal cortex to more automatic and evolutionarily ancient structures as training progresses.

For Soldiers and TSUs to take decisive action in the continuous, simultaneous combinations of all tasks, they will often have to make decisions almost immediately, which heightens the importance of highly effective training aimed at developing effective rapid decision-making skills in every area of capability. Soldiers and TSUs thus must be given both the human abilities required to make decisions that are as nearly optimal as possible and the materiel required to execute those decisions in a timely manner. From a human dimension perspective, Soldiers and TSUs must receive the training required to achieve effective near-optimal decision making in a rapid and intuitive way, as measured using robust and well-validated measures of effectiveness, prior to entering the operations area. For stability tasks, cultural knowledge and awareness are critical elements of this process. Leaders, in particular, need honed decision-making processes for decisive courses of action and dominant execution using both rapid intuitive mechanisms developed during training and slower deliberative processes that can be leveled at the novel and unanticipated situations that always arise in operational environments.

In addition to training, advances in materiel systems would also benefit the Soldier and TSU leadership. Advanced information systems (including sensors) are needed to provide actionable information to a decision-maker. Additionally, cognitive decision aids (broadly including cognitive agents, decision aids, expert systems, augmented cognition, and the like) would be of benefit for improving a leader's decision-making capabilities.

Better decision-making and the ability to execute those decisions effectively will always be a critical factor in decisive combat engagements. But they are also critical to enhanced survivability, efficient accomplishment of stability tasks, and improved competency in wide area security. The TSU needs to develop complex decision-making skills for the entire range of military operations, including the ability to succeed in nonlethal situations that involve close contact with civilian populations. Decision-making is necessarily pushed down to the lowest tactical level to perform assigned tasks based on specific information requirements. The scope of TSU tasks goes well beyond offensive and defensive combat operations to include tasks that require varying aspects of DIME (diplomatic, information, military, and economic) and PMESI (political, military, economic, social, and infrastructure) information. To be effective on the streets and in the villages with noncombatants and potential threats alike, the TSU must have cultural awareness and understanding, as well as language skills. The TSU must also have ready access to the cultural and social relationship knowledge of the mission space gained by other units operating in that area previously, in adjoining areas, and in the region.

Three Levels of Situational Awareness

Of particular importance to decision-making and execution from a perspective that addresses both the human and materiel dimensions is the need for personal and shared situational understanding (also called enhanced situational awareness). A simple definition for situational understanding is "the perception of the elements in the environment within a volume of time and space, the comprehension of their meaning, and the projection of their status in the near future" (Endsley et al., 1998). Situational understanding for a TSU can also be viewed as the coordinated perception of change in the operating environment by the individual Soldiers, each of whom sees only a portion of that environment. Full situational understanding requires three levels of situational awareness, characterized as follows:

Level 1 situational awareness (perception) is the Soldier's or TSU's perception of disaggregate elements of information. For the Soldier or TSU, this includes but is not limited to seeing all aspects of the operational environment, including physical surroundings, enemy forces, noncombatants, friendly forces, terrain, and weather. An example is the perception of sensor images of civilians meeting with insurgents. Level 1 questions might be as simple as "Where am I?", "Where are my buddies?", and "Where is the enemy?" More complex questions might be "Which friendly units have been detected by the enemy?" and "Which enemy units are currently firing/applying weapons?"

Level 2 situational awareness is the level at which the Soldier or TSU first gains situational understanding. It is achieved when the Soldier or TSU combines, interprets, stores, and retains the Level 1 information. The interpretation of the information is achieved using personal cognitive filters and fusion algorithms developed through training, combat experiences, and perhaps intuition. It includes integrating information received and determining the relationship among pieces of information and the relevance of the individual pieces to a desired end state. As an example, using the images of civilians interacting with insurgents, along with other information, one determines if the civilians are friendly, neutral, or antagonistic toward the insurgents. A Level 2 question might be: "What is this enemy unit's objective?"

Level 3 situational awareness is an extension of understanding. It is reached by using understanding to project possible future events and to anticipate their outcomes. Soldiers and TSUs project future events as an outcome of the current understanding of the operational environment joined together with anticipated events that will impact the desired end state. An example of Level 3 situational awareness would be the ability of a TSU to determine if civilians will be honest and open in their discourses with U.S. forces. A Level 3 question might be: "What do you expect this enemy unit to do in the next 10 minutes?"

For Level 1 capability, materiel dimension solutions (e.g., geolocation systems and information technology systems) may play a significant role, while for Levels 2 and 3 human dimension solutions (e.g., training, enhanced memory, enhanced cognitive performance) will often have a higher priority. Coordination and communication among the Soldiers within a TSU or between TSUs engaged in a shared mission or task typically require and add to Level 2 and 3 situational awareness, thereby increasing situational understanding.

Network Integration

Currently, when a TSU leaves a forward operating base (FOB) or disembarks from a vehicle, it has very limited access to technology for command decision tasks such as communicating, developing situational understanding, and understanding the human terrain. A squad leader's communications system provides bandwidth rates in the tens of kilobits per second—a far cry from the multiple megabyte rates available within a FOB. Sand tables and paper maps support mission rehearsal and execution. Sensing during a mission is primarily dependent on the eyes and ears of members of the TSU. These shortcomings prevent TSUs and Soldiers from achieving optimal performance in making and executing personal and team decisions.

TSUs must be integrated into the Army network, with the network being pushed down to the individual dismounted Soldier. They must have enhanced capability in functional areas such as mission command, intelligence, fires, and mission planning/execution. For example, there should be timely, relevant information on location of friendly assets, the identification and location of enemy forces and equipment, the identification and location of noncombatants, and the ability to document and communicate this information to each other and higher echelons. Information should be streamlined to minimize risk of information overload, and it must be timely to ensure that TSUs are not surprised in tactical situations. More responsive reach-back capabilities are also needed to enable synchronized employment of supporting weapons platforms such as mortars and artillery.

The integration of Soldiers and TSUs into the Army network would enhance capabilities in all areas, not just decision-making and execution. To provide the TSU and individual Soldier these enhanced capabilities, advances are needed in communications, information, and socio-cognitive networks.

Communications Networks

Communications at the TSU level—among the Soldiers in the TSU, with robotic systems within their operational environment (systems either organic to the TSU or attached to higher echelons), and between the TSU and leaders at higher echelons—are very poor with respect to range (especially in urban areas) and bandwidth rates (associated with the limited frequencies available at the small unit level). The TSU lacks the capability to send and receive secure data, voice, and streaming video at adequate ranges and with sufficient reliability.

For communications networks, advances are needed in hardware, frequency spectrum (particularly for bandwidth rates), and user interfaces. The Army is attempting to address these needs with the Nett Warrior Program, and with experiments using smart phones, thereby leveraging Soldiers' familiarity and comfort level with personal wireless technologies. However, the Nett Warrior Program is limited by low bandwidth, and the latter effort is dependent on commercial networks. High-bandwidth communications networks are needed that can operate in austere locations, in complex terrain (e.g., urban or mountainous), in all weather, and under day and night conditions. Night operations require communications devices that should not compromise one's location and should be usable with night vision devices.

Current Army initiatives emphasize the "Network" and the "Forward Edge," which translates into a need, at the TSU, for secure, reliable, and sufficient bandwidth rate to support intra-squad communication, the operation of organic or attached equipment, and links to higher and adjacent units, as well as elements of the local population. This demand for bandwidth is further complicated by the likelihood that the squad will be operating in a joint, coalition, and expeditionary environment, often in difficult terrain and often austere from a communications infrastructure perspective. Given that available bandwidth is usually saturated in the operational environment of a dispersed small unit deployment, it is important to provide dynamic, situation-based management of this critical resource.

Information Networks

Information exchange (especially for digital images and streaming video) is currently very poor at the TSU level. Bandwidth rate is one issue. Another is that operation tempo (OPTEMPO) does not give TSUs time to download, evaluate, and make judgments on available information; that is, they very easily reach information overload. The need for information varies with mission. While a single image will suffice during an operation, video is needed during planning. TSUs and Soldiers would benefit from advances in dynamic information networks that enhance information exchange and information assessment capabilities.

Technologies and procedures are needed to ensure TSUs have access to information networks. Even if communications networks are enabled, there is no guarantee that TSUs and Soldiers will have access to needed information. For example, as GEN Peter Chiarelli stated in a network-centric warfare conference on January 23, 2008, "Information is firewalled by the bureaucracy. Commanders are unable to get the information they need because of bureaucratic obstacles." GEN Chiarelli went on to say that insurgents lack the sophisticated equipment of the U.S. military, yet they have become highly adaptable foes merely by using cell phones, video cameras, Internet access, and e-mail (Matthews, 2008). An example of a bureaucratic obstacle is operational security, where classified information from a non-Army source is not shared with TSU Soldiers because of their minimal security clearances and the possibility they may be operating with uncleared personnel (e.g., host nation forces).

TSUs need access to a variety of information networks (including databases) and a variety of sensors (using various mediums, situated on the ground or airborne, manned and unmanned). Capabilities should provide access to both internal (assigned to the TSU) and external (e.g., unmanned aerial vehicles assigned to battalion headquarters) information sources. One of the most critical information needs is knowing the identification, location, and tracking of friendly, enemy, and noncombatant personnel, especially in cluttered urban environments where Global Positioning System (GPS) signals are weakened or completely blocked. Another is the ability to sense through walls or on the other side of obstacles.

Socio-Cognitive Networks

When interacting with noncombatants in irregular warfare counterinsurgency operations, TSUs would benefit from a wide array of socio-cognitive networks and biometric tools. For example, when entering a village, a TSU could use advances in dynamic socio-cognitive networks to: (a) identify a person's community, (b) identify a person's association with overlapping communities, (c) identify and interact with local leaders, and (d) visualize a leader's/person's connections. Within their FOB, TSUs have access to such socio-cognitive networks (e.g., the Tactical Ground Reporting Network) and biometric systems. However, once they depart the FOB, links to these tools are severed. Access to such tools will enable better situational understanding of the human terrain.

In combat situations, these networks should support the Soldier's and TSU's ability to rapidly shape the operational environment before engagements by exploiting every aspect of the populace for its advantage in decreasing the threat from noncombatants, including minimizing collateral damage or loss of noncombatants.

MILITARY EFFECTS

In the context of the Army role in unified land operations, military effects include much more than the traditional combat capability to produce lethal effects on an opposing force (see Appendix C). The dismounted TSU and Soldier also need nonlethal capabilities to counter, control, disarm, or disperse individuals who may or may not be a potentially lethal threat. Military effects also encompass the ways to influence and support people and communities, in order to succeed in stability tasks.

Lethality

The following statement on squad lethality is taken from an article entitled "The Infantry Squad," written by LTG Robert Brown when he was Commander of the Maneuver Center of Excellence:

> The ability to find, fix, and finish the enemy is paramount to any tactical formation. We must maintain it and improve upon it. The squad's weapons must complement each other and give the squad the capability to use both precision direct fires and devastating area fires. Ammunition should kill or incapacitate an armored enemy as well as an insurgent without body armor. We must also maintain and improve the squad's capability to deliver high-explosive counter-defilade fires against an entrenched enemy.
>
> (Brown, 2011, p.9)

In this statement, the key words for defining the lethal effects capability desired for a dismounted TSU are "weapons must complement each other," "use both precision direct fires and devastating area fires," "ammunition should kill or incapacitate" and "deliver high-explosive counter-defilade fires."

The current TSU's complement of weapons makes the TSU quite lethal. Of course, there is always room for enhancing weapon capabilities—to include increased range and terminal ballistics performance—and shooter performance. The addition to the squad weapons inventory of the XM-25 rifle, with its airburst rounds, will significantly enhance the TSU's lethality. TSU lethality can be further enhanced with the following improvements to offensive and defensive operations.

TSUs must be able to find, fix, and engage an enemy at their choosing. They need the ability to initiate contact rather than be surprised and have to reach to contact. TSUs must be able to quickly synchronize organic and supporting Army fires (e.g., heavy machine gun, mortar, artillery), as well as joint fires. TSUs must have the capability to employ precision targeting—e.g., GPS-guided mortar rounds that allow for control prior to firing and during the flight of the round. Unique to defensive operations is the TSU's need for lethal capability against tactical improvised unmanned aerial vehicles.

Overall TSU lethality is greatly dependent on the individual Soldier's weapon performance. Cross-training a Soldier to be proficient in a number of weapons provides the TSU leader with increased flexibility in combat lethality as TSU members are injured. As budgets become tighter and training ammunition is less available, it will be difficult to maintain individual Soldier proficiencies on multiple weapons.

Over the past 10 years, the Army's engagements in Iraq and Afghanistan have highlighted the importance of the dismounted Soldier in unified land operations. These engagements have also highlighted the many shortcomings that still exist in making the Soldier dominant (giving the Soldier decisive overmatch) across the range of military operations. Although this section focuses primarily on the dismounted infantryman, the infantry squad, infantry platoon, infantry company, and infantry battalion, one should never forget that all Soldiers, regardless of operational function, have similar requirements in the areas of lethal and nonlethal effects and protection. Ensuring that the tip of the spear is dominant will also better enable all forces in the operational environment.

The equipment of an infantry Soldier has a generic component, including items such as uniforms and body armor that apply to all of the Soldiers in an operational environment, and an assignment-specific component, which depends on the Soldier's position within a unit and the type of unit (e.g., heavy infantry, infantry, rangers, or Stryker infantry). For example, by the modified table of organization and equipment, infantry rifle squads are configured the same regardless of parent organization—that is, whether they are in a light infantry, infantry, air assault, airborne, or ranger unit. Figure 2-2 depicts the generic dismounted infantry squad currently found in all types of units.

The primary difference in the squad's lethal component exists at the platoon level, where the rifle squad is supported by either two machine gunners at the platoon headquarters in light infantry organizations, three machine gunners in the weapons squad of a ranger platoon, or two machine gunners and two antiarmor Javelin gunners in infantry, airborne, and air assault platoons.

The lethal component differences are even further amplified when one compares the direct fire support available to heavy infantry squads from supporting Bradley Fighting Vehicles and Abrams Tanks and for Stryker infantry squads with the direct fire support from their infantry carrier, anti-armor tube-launched, optically tracked, wire-guided missile carrier, and mobile protected gun. Obviously, the mobility and load carrying capabilities of the squad also vary greatly by the type of unit.

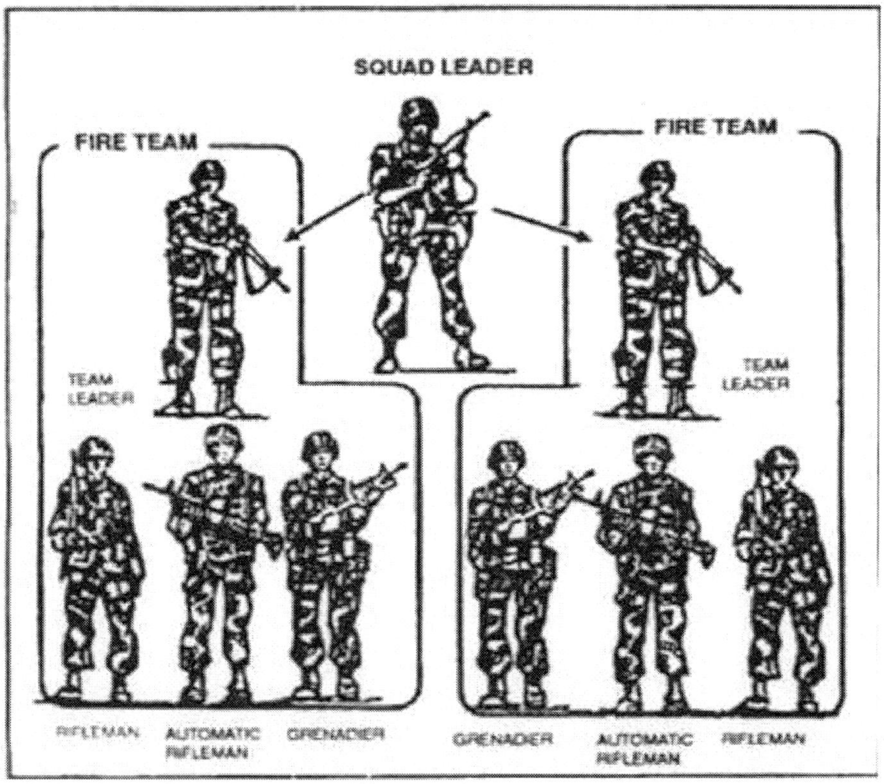

FIGURE 2-2. Generic U.S. Army rifle squad. SOURCE: U.S. Army, 1992.

To be decisive in stability and other operations short of deadly combat, dismounted TSUs also need less-than-lethal alternatives to lethal weapons. Such nonlethal weapons can give Soldiers and TSUs more engagement options and prevent the escalation of tense situations in stabilization operations.

Stability and Humanitarian Effects

There are many capabilities that are unique to or have greater importance in stability and civil-military operations (Chait et al., 2006; Chait et al., 2007). These include but are not limited to the following examples.

- Wide area security includes the wide area security tasks discussed earlier in this chapter.
- Patrols for stability tasks. Often these patrols are not combat patrols but rather humanitarian, military police, and presence patrols. TSUs must be able to quickly transition, both mentally and physically, from being a combat patrol to a humanitarian or police patrol.

- Checkpoint security. To allow traffic to flow through checkpoints securely yet efficiently, which supports the legitimacy of governance and the stability objective of socioeconomic development, TSUs need the capability to screen people and equipment quickly, yet be able to identify individuals of interest and weapons or contraband with a high probability of detection and low rate of false positives. Sensors and information systems are needed that can: (1) quickly compare visible physiological markers against an intelligence database maintained at higher echelon, and (2) identify molecular and other signatures of hidden or disguised explosives, contraband, and weapons.
- Communicating and information sharing with non-U.S. security forces and non-military personnel, including personnel from nongovernmental organizations (NGOs). TSUs must be able to rapidly communicate and share information with security forces from allied nations and from the host nation local security forces (both military and police). TSUs also need to communicate and share information with non-military personnel, especially representatives from the U.S. Department of State, foreign governments, and NGOs (e.g., the International Red Cross).
- Actions based on situational understanding of State Department and NGO operations. The actions of the TSU must be aligned with and support State Department operations (especially the U.S. Agency for International Development) and NGOs. This is especially important if the TSU must transition to combat tasks in the midst of a primarily stability/humanitarian operation.
- Actions that require enhanced cultural awareness. For situational understanding of the needs and perspectives of the local populace, an increased cultural awareness is needed by both Soldiers and TSUs, so as not to commit, for instance, a *faux pas* that negates weeks or even months of hard work in winning the support of the local populace.
- Increased demand for data collection. To support the commander's assessment of stability operations, a significant burden may be placed on the TSU to collect data to support assessments in the following four areas:
    - *Security* is the protection from threats and other activities of insurgent, terrorist, criminal, nationalist, ethnic, and extremist groups.
    - *Governance* is the collective process of decision-making and the process by which decisions are implemented (or not implemented). It may be analyzed by three components: process, participation and accountability.
    - *Rule-of-law* is dispute resolution as it applies to person-to-person, person-to-group, and group-to-group disputes. Rule-of-law may include familiar systems such as a constitution and national laws; local district or village laws; and courts, judges

and police forces; as well as systems unfamiliar to most Americans, such as religious laws (e.g., Sharia laws).

At the dismounted TSU level, as well as at the theater-wide level of Army operations, socioeconomic support capability includes actions that build capacity of social and economic institutions so they may withstand and diminish threats such as those identified above under "security." Examples of this capability include establishing governing institutions, improving the existing transportation infrastructure, providing basic needs (water, electricity, sewage, etc.), expanding the existing education infrastructure, improving access to medical facilities, and providing high-impact economic (agriculture and industry) assistance. Related objectives include reducing illicit economic activities that undermine stability objectives, such as local or national corruption and illegal or harmful economic activities. Examples of the latter in recent Army theaters of operation include interdicting cultivation of opium poppies and the processing/distribution of narcotics in Afghanistan.

MANEUVERABILITY

Tactical maneuverability (combination of mobility and agility) is difficult to achieve in complex, austere, and harsh terrains and at a high OPTEMPO. To effectively close with and neutralize the enemy utilizing fire and maneuver, mobility for the Soldier and TSU must be equal to or better than that of adversaries. Survivability focused on heavy personal armor will reduce mobility, so survivability ensembles must allow for adversary-competitive mobility, while keeping casualties within strategic expectations.

TSU and Soldier maneuverability needs vary with roles, missions, and phases of a mission. For example, dismounted rifle TSUs (those that close with and neutralize the enemy) will require more maneuverability than the heavy weapons TSUs (e.g., those in a heavy weapons platoon) (HQDA, 2007). Additionally, TSUs augmented by heavy weapons—such things as heavy machine guns, mortars, anti-tank weapons, and associated ammunition—have greater need for improved mobility rather than agility. With regard to phases of missions, TSUs carry a maximum load to assembly areas, a smaller load to pre-assault positions, and finally their combat load during the assault. During current operations, the unloaded equipment is secured by other parent organizational elements. In future operations, autonomous ground vehicles may be available for carrying the loads, but the TSU may still have to offload the equipment, swap the equipment out during each phase of the operation, and provide for security.

TSUs also need better maneuverability in complex terrain (e.g., urban, mountainous, and jungle). As mentioned earlier for urban operations, TSUs must not be constrained by ground-level doors and windows for assaulting a building nor by stairwells for vertical movement within a building.

SUSTAINABILITY

Small units will be called upon to operate independently for extended periods of time with increasing OPTEMPO. For the Soldier and TSU, this translates into sustainability needs for power and energy less tethered to the logistics base. It also signals the need for innovations in training both on the ground and in the Army's traditional training centers (or schoolhouses).

Power and Energy

Energy is a ubiquitous quantity, and the term is often used interchangeably with "power," which is the rate at which energy is used. By the first law of thermodynamics (conservation of energy), energy cannot be created or destroyed, only changed from one form to another. In a sense, it is the "currency of the universe" in that everything we see, measure, construct, and do has an energy budget associated with it. As a result, there is a continuing search for dense forms of energy that can be readily applied to the insatiable appetite of a growing world population. Energy can be extracted from its storage in the atomic nucleus, chemical bonds, or gravitational field and from energy sources such as the wind and solar radiation (which ultimately derive from nuclear energy sources in the Sun).

Two prior National Research Council (NRC) studies considered power and energy for the dismounted Soldier exclusively (NRC, 1997; 2004). These studies concluded that power reductions and conservation must be part of the overall solution to meet Soldier's needs, but in ensuing years energy and power demands for the dismounted Soldier have only increased as the numbers and variety of electronics in his equipment have proliferated. This study, unlike the earlier studies, examines the needs for power and energy within the overall context of ensuring that dismounted TSUs have decisive overmatch through superior capabilities in the areas of situational understanding, military effects , maneuverability, supportability, and survivability.

Why Energy is a Problem

The focus of this study is the individual solider and how to make him/her overwhelmingly superior to any adversary. The revolution in digital technology has made it possible to equip the Soldier with unprecedented capability such as real time situational awareness through computer displays that overlay data on maps showing the location of friend and foe, local Internet-like capability, and personal weapons that use electronic systems to enhance lethality. All of these capabilities are powered by local energy sources carried by the Soldier. If one looks at the Soldier on today's battlefields in Afghanistan and Iraq, the image is one of a grossly overloaded Soldier in the hot desert sun, struggling with total loads sometimes exceeding 100 pounds or more for extended missions (Figure 2-3). A substantial fraction of that load is associated with the energy supply needed to power the Soldier in the form of food, to power his lethal equipment in the form of explosives or propellants, and in the form of batteries to drive the ever-expanding array of electronic tools designed to improve his fighting skills and make him more decisive.

FIGURE 2-3. The modern warrior with combat load during dismounted operation in Afghanistan. SOURCE: CALL, 2003.

For a dismounted mission, Soldiers must carry all of their energy in various storage formats or rely on others to provide them with timely resupply. In any case, it requires expenditure of energy to construct suitable energy storage devices that dismounted Soldier will use in addition to requiring energy to transport resupply to them. These expenditures translate to monetary costs to produce energy storage units, transport the units to the Soldier, and store them in theater. To a large extent, these costs drive what is available. Since Soldiers are limited in what they can carry, improving the density of energy storage media is a primary concern. Different modes of energy storage can be compared using a common measure of energy density such as watt-hours per kilogram. A concern for efficiency in developing and distributing energy resources also leads to the need for an energy cost metric, such as dollars per watt-hour or dollars per gallon of logistic fuel, delivered to the Soldier.

How Much Energy Is Needed as a Function of Mission?

For each mission, there is an associated energy requirement to carry out that mission. It is instructive to examine the source of the total mass associated with the energy needed to carry out a particular mission. The energy sources currently in use are ultimately traceable to energy stored in chemical bonds, which sets an upper limit on how much energy can be extracted from a unit of mass.

In terms of today's technology, Table 2-1 illustrates the energy formats and the amount of energy required for a 72-hour mission, assuming no resupply is available. If a Soldier's load for this mission is 100 pounds, roughly one-third of the mass is associated with his energy supply. The remainder of the weight is associated primarily with body armor, weapons, and the weight of electronic devices such as radios.

TABLE 2-1. Energy Formats and Amounts of Energy Required for a 72-Hour Mission

Energy Format	Function	Number of units	Approximate energy W-h (MJ)	Approximate mass, kg (lbs.)	Energy density, W-h/kg
Food (MREs)	Power the Soldier	9 meals	10,800 (30)	6.1 (13.5)	1,770
Ammunition (30 rounds) and 2 grenades	Lethal agents	~1 kg TNT explosive equivalent	1,278 (4.6)	1.8 (4)	710
Batteries, average draw 9.17 W	Power equipment ensemble	7 battery types 70 total	660 (2.4)	7.2 (16)	92

In evaluating capability needs, the committee concentrated primarily on energy needed to power the electronic items carried by the Soldier which make up a large part of the total mass associated with the Soldier's energy supply. Items such as an exoskeleton, which are still in early research and development (R&D) stages, will have their own power systems.

SURVIVABILITY

Survivability includes needs related to protection, which runs the gamut from individual Soldier protection to small-unit force protection to layers of protection external to the TSU. For both TSU and individual Soldier protection; there is insufficient force protection to ensure the highest degree of survivability across the entire range of military operations.

A challenge is to balance protection with other capability needs, such as maneuverability and military effects. The protection goal is not to focus solely on reducing damage but to focus on significantly reducing the threat's ability to detect, attack, or hit the Soldier and TSU. For example, the threat's ability to attack can be reduced by detecting and neutralizing threat personnel and weapons before the threat can engage the TSU. Protection should not degrade the TSU and

Soldier from maintaining momentum. Finally, protection should be considered with respect to kinetic (e.g., bullet, fragmentation, blast, chemical/biological) and nonkinetic (e.g., nonlethal weapons, overheating, musculoskeletal injury) events.

Individual Soldier Protection

With respect to kinetic events, most Soldier protection has been focused on body armor and helmets, which have been proven to reduce significantly the probability of damage to the torso and head, respectively. However, protection of arms, legs, and pelvic areas is also a dire need. A September 20, 2011, article in *USA Today* reported that through July (58 percent of the year), 134 warfighters had lost limbs in combat in 2011, which is 78 percent of the 171 amputations in 2010. In addition, there have been 79 cases of multiple amputations in 2011, more than any previous year. When a limb is lost or amputation is required, there is often damage to the lungs, kidneys, and liver from massive blood loss and shock. Infections are severe because sand peppers non-armored areas and fungi penetrate deeply into body wounds. 90 of the wounded troops had to deal with lost genitals from blasts (Zoroya, 2011). Injuries in this last category create significant psychological issues in addition to the physiological damage.

Kinetic events also contribute to brain injuries, especially concussions, which are highly correlated with subsequent post traumatic stress disorder. About 15-20 percent of all Soldiers sustain concussion during deployments.[4] The number of traumatic brain injuries needs to be reduced.

The best protection against a kinetic injury event is to prevent the event from occurring at all. Capability enhancements in sensing, individual situational awareness, and shared situational awareness can prevent the TSU from being surprised and allow it to maintain the initiative. Stand-off sensing (e.g., sense-through-the wall, remote sensing) and engagement capabilities will assist in keeping Soldiers out of harm's way, thus reducing the probability of combat injuries.

Soldiers and TSUs are also susceptible to combat injuries other than from kinetic hits. As Soldiers and TSUs become more dependent on electronic-based systems (e.g., communications networks, information networks, night vision, geo-location systems), they become more susceptible to electronic warfare and directed energy weapons (high power microwaves and high energy lasers). Directed energy weapons, such as high energy lasers, also pose a threat to a Soldier's eyes.

The Soldier's load is a serious issue for unit survivability as well as for maneuverability (agility), as it creates many noncombat injuries. In fact, the load weight is the largest contributor to noncombat injuries—24 percent of medical evacuations from Operation Iraqi Freedom (Iraq operations) and Operation Enduring Freedom (Afghanistan operations) were noncombat musculoskeletal

[4]COL Gaston P. Bathalon, Commander, U.S. Army Research Institute of Environmental Medicine, "The Soldier as a Decisive Weapon: USAMRMC Soldier Focused Research," presentation to the Board on Army Science and Technology, February 15, 2011.

injuries.[5] As another example of load effects, Soldiers overheat in hot environments, but the problem is made significantly worse by heavy loads, which interfere with normal heat dissipation, and by batteries as they give off heat during use. For these and many other reasons, there is a critical need to reduce Soldier load to improve small-unit survivability, as well as to solve maneuverability and sustainability problems.

Operational deployments, combat, and high OPTEMPO activities lead to physiological strain and fatigue. As one metric, Soldiers suffer weight loss (20 percent of combat Soldier suffer more than a 5 percent weight loss) and performance deficits due to unmet nutritional requirements (both food and water).[6]

Finally, the threat of chemical/biological/nuclear weapons cannot be ignored. Given the current combat loads and added protective gear and clothing, dismounted Soldiers cannot react quickly to unconventional warfare attacks. The problem is even more severe in extreme weather environments.

TSU Protection

At the TSU level, force protection is the primary focus. Collective capabilities are needed to support the protection of both the TSU and its individual members. For example, assuming a TSU composed of nine Soldiers, the TSU leader can enhance protection from load-based environmental injuries by sharing tasks (e.g., sensing, heavy physical missions) and redistributing weapons and load (e.g., ammunition). However, more needs to be done to enhance protection for the TSU as a whole. Enhanced shared situational awareness (unit-level situational understanding) is one example, and it can be enabled with network improvements described in the Materiel Dimension section below. Enhanced TSU mobility will also improve protection. A third example of a unit-level capability is gunshot location detection, for locating trained snipers as well as untrained sharpshooters.

During dismounted operations, TSUs currently lack capability to accurately detect, at safe distances (up to 100 m), changes in surface and subsurface (3 to 6 inches below the surface) conditions. This capability could be useful in identifying and avoiding land mines and improvised explosive device threats.

Although it was not directly mentioned in the reviewed Army documents or in discussions during committee site visits to Army bases and laboratories, a potential threat to dismounted TSUs that the committee believes is emerging and should be addressed is that of tactical-level air systems, such as autonomous killer drones (Finn, 2011). An even nearer term, low-tech threat could come from improvised unmanned aerial vehicles (e.g., remote-control model airplanes

---

[5]COL Gaston P. Bathalon, Commander, U.S. Army Research Institute of Environmental Medicine, "The Soldier as a Decisive Weapon: USAMRMC Soldier Focused Research," presentation to the Board on Army Science and Technology, February 15, 2011.

[6]Ibid.

carrying explosive payloads). For example, it would be extremely difficult to shoot down a small, commercially available drone air vehicle fitted with an improvised explosive device, using only the personal weapons (M16, M4, M203, and M249 machine gun) currently carried by a rifle squad. Improvements in both materiel and human dimensions are needed to enhance a TSU's capability to sense and neutralize such threats before they reach their target.

In urban offensive operations, dismounted TSUs still depend on entering buildings through doors and windows—normally on the first floor. Once inside a building, Soldiers are confined to moving to different floors via stairwells. These movements give an adversary the advantage in terms of surprise and knowing the TSU's avenues of advance. Capabilities are needed for a TSU to maintain surprise and the initiative in urban operations.

In defense in open terrain, TSUs still depends on entrenching tools to dig in for protection. Dismounted TSUs need the capability to more quickly establish defensive positions in open area operations (AUSA, 2011).

Layers of Protection External to the TSU

Supporting counter rocket and mortar systems will help protect Soldiers and TSUs from enemy indirect fires. However, fratricide is an unfortunate result of mistaken identity. Without integrated identification, location, and tracking of friendly forces across all services, the TSU is susceptible to fratricide.

The majority of wide area security missions will be conducted jointly with military and nonmilitary activities external to the TSU. Where and when appropriate, the TSU must be knowledgeable of and integrated into these activities, especially when the TSU is conducting stability tasks that could unexpectedly turn into lethal combat.

CURRENT OPERATIONAL WEAKNESSES

Weaknesses in current dismounted Soldier operations provide insights into ways that the decisiveness of the TSU and individual Soldier can be increased. Following are examples of current capability weaknesses in dismounted TSU operations that were identified during committee member interactions with troops and officers in units recently returned from deployment.[7]

- Once a TSU leaves the FOB or a vehicle, its access to tactical and socio-cognitive information is severely limited.
- It is very difficult to make quick changes in a TSU's line of advance when it is engaged in a mission (e.g., moving the unit within minutes from one

---

[7]Committee members interviewed U.S. Army commissioned and noncommissioned officers at Meeting 1, held at Fort Benning, GA, July 12-14, 2011.

ridge line to a parallel ridge line across a valley), due to its limited tactical mobility while dismounted.

- In determining the load required to accomplish a mission, providing protection and survivability is nearly always in conflict with preserving the individual Soldier's physical ability to be mobile and agile enough to fight the enemy.
- Soldiers are susceptible to both the physical and mental harm caused by the harsh conditions of combat, the effects of direct and indirect fire, the physical environment, psychological combat stress, and even personal issues at home.

Human Dimension Issues

The TSU must have the ability to operate effectively (operate as planned) in extreme environments, e.g., at high altitudes. To make best use of the weight carried, the TSU members must increase their lethality through increasing kinetic combat skills (specifically marksmanship) during basic and advanced training. They must exhibit adaptability in the "three block war" by being able to shift rapidly between kinetic and nonkinetic operations and to adjust to the global visibility of local operations, where each Soldier is a "strategic corporal."[8] Finally, the TSU must have the emotional and mental resilience to withstand and adapt to rapidly changing conditions.

To be dominant within its assigned area of operation, a TSU must be able to deliver, or cause to be delivered, lethal and nonlethal effects against threats, with ranges and accuracies greater than the threats; to be able to discern threats from friends and noncombatants, again, with greater ranges to and accuracies of identification greater than the threats; and to be able to outmaneuver the threats. It must be able to achieve these military effects for the full duration of any assignment. Within their assigned area of operation, the Soldiers of this squad-level unit must be able to move with the confidence that they have awareness of the location and intent of physical and personnel threats or humanitarian needs; access to the resources, training, and physical protection required to carry out the assigned mission; and the background knowledge and skills required to accomplish that mission.

Performance Degradation Factors

Second only to unit design in maximizing TSU performance is the physiological performance of TSU members. Significant ongoing Army research suggests that squad members during recent deployments were often operating at

[8]The strategic corporal is the notion that leadership in complex, rapidly evolving mission environments devolves lower and lower down the chain of command to better exploit time-critical information in the decision-making process, ultimately landing on the corporal, the lowest ranking noncommissioned officer,

low levels of physiological performance when they were executing tasks in operational environments (see, for example, Miller et al., 2011).[9] Any change to Army doctrine and technology that can improve the average physiological performance of squad members would have a profound effect on the performance of TSUs. Factors that are known to compromise Soldier performance include sleep loss, Soldier load, other physical stressors, and emotional and psychological stressors and resiliency.

Sleep Loss. Depending on the individual, physiological performance is compromised to a varying degree by long periods without sleep and with only brief periods for recovery from sleep deprivation. While the size of the Army and its mission largely dictate the OPTEMPO faced by TSUs, it seems clear that insufficient attention is paid to the predictable performance decrements caused by sleep deprivation. Even when it is not possible to avoid longer missions that preclude sleep, it is possible to model and understand the decrement in performance, and thus the loss of squad decisiveness, that inevitably accrues as squads become progressively sleep-deprived. Understanding sleep deprivation, how slowly performance recovers from sleep deprivation, and how to predict a squad's loss of decisiveness should be a critical feature of mission planning to ensure decisive overmatch—although this is clearly not the case today.

Soldier Load. Heavy combat loads degrade mobility in combat, reduce ability to maneuver for advantage through accelerated physical fatigue, degrade cognitive performance, and contribute significantly to both noncombat casualty evacuations and career-ending disabilities. Data from the U.S. Army Research Institute of Environmental Medicine showed that 24 percent of medical evacuations from Iraq and Afghanistan were for noncombat musculoskeletal injuries; 72 percent of medical discharges were due to chronic musculoskeletal injuries.[10] Lightening the Soldier's load is critical. However, given Soldiers' and small-unit leaders' load-carrying behaviors, the challenge is more than just reducing the weight of the assigned, required, and expected individual equipment carried today. Defense materiel vendors as well as Army laboratories and Army research, development and engineering centers have shown in technology demonstrations that the weight savings from disciplined iterations of very focused cycles of engineering design, build, and evaluation are not the answer, only on the order of single-digit pounds. Addressing Soldier load will therefore require looking for other opportunities for equipment integration and load reduction, including offloading to a carrier system.

---

[9]More broadly, during the committee's September 15-16, 2011, visit to the Natick Soldier Systems Center and to the U.S. Army Research Institute of Environmental Medicine, also at Natick, research staff described surveys and casualty analyses that indicate warfighters in Operation Iraqi Freedom and Operation Enduring Freedom have been hampered by suboptimum nutrition; progressive, chronic musculoskeletal injuries; and altitude and heat stresses.

[10]COL Gaston P. Bathalon, Commander, U.S. Army Research Institute of Environmental Medicine, "The Soldier as a Decisive Weapon: USAMRMC Soldier Focused Research," presentation to the Board on Army Science and Technology, February 15, 2011.

Other Physical Stressors. Extreme heat, extreme cold, high elevations, and heavy loads all compromise physiological performance. The degree of this compromise can be predicted and considered by mission planners. Ideally, the TSU should be ready for each mission as "game day," without physiological lapses that sap performance. When this is not possible, upper echelon decision-makers must understand the compromises they make when they select units at different levels of physiological performance for particular missions.

Emotional and Psychological Factors. Soldiers in a dismounted TSU must be emotionally and psychologically ready to perform in their assigned roles if that unit is to achieve its potential for dominance. Traditionally, Soldiers rated as deployable by their medical officers have been assumed to be combat ready, but there seems to be growing anecdotal evidence that this is not always the case. The efficacy of many deployable solders may be compromised by the emotional stresses of the operational environment, and this inevitably detracts from the ability of their units to achieve decisive action. While there has been some progress in this area, and resiliency training has been adopted to some degree (for example, through the Army Center for Enhanced Performance), modern human factors tools have not yet been fully employed in this area.[11] Psychological tools exist for assessing the emotional efficacy of individual Soldiers and for quantifying the loss of efficacy that occurs over a Soldier's deployment cycle. Tools also exist for enhancing resiliency and improving selection when TSUs are constructed and maintained. A focused and funded program of relatively inexpensive research to target this area might well find that very significant performance improvements are possible in this domain as well. Those emotionally unsuited for TSU assignments must not be allowed to be distractions.

Manning and Training

The current manning and training structure of the U.S. Army squad both defines and enables the training, tactics, and procedures that the Army executes. The mix of personalities, the experience of squad members, the network of trust, the resiliency of individual Soldiers, and the number of individuals within the squad have at least as much effect on operational effectiveness as does the hardware these individuals carry into the operational environment. But in fact, the effects of each of these critical features on performance are largely unknown and unstudied. The assumption that the modern squad size and structure is efficient thus remains almost entirely untested at an objective level. Would larger squads like those used by the Marines be more effective per Soldier? Would a shift towards more experience, by increasing the proportion of second and third term enlistments in TSUs (and thus an Army-wide shift towards longer tenures), yield

[11]LTC Carl Ohlson, Academy Professor and Director, Center for Enhanced Performance, U.S. Military Academy at West Point, "Army Center for Enhanced Performance Overview," presentation to the Board on Army Science and Technology, February 16, 2011.

a more effective unit per dollar spent? Could changes in training that are known to alter unit cohesion or interpersonal trust also influence unit effectiveness? In any objective sense, the answers to these questions remain unknown.

What is known, however, is that TSUs have not been as dominant in recent operational environments as the Army would like them to be. Also known is that alternative squad designs are possible. If performance metrics were available, it would be possible to assess the costs and benefits of alternative designs and thus to ask whether a single squad design alternative is optimal or whether multiple reconfigurable squad designs would provide superior operational performance, given the anticipated range of military operations. The TSU of the future may well have to aggregate and disaggregate as required. This consideration seems particularly relevant as Army doctrine shifts from fixed, traditional divisions of the last century to the more nimble and dynamic BCTs of today—and it may shift back.

TSU Organization

Today's nominal U.S. Army squad consists of six Soldiers of low rank (E4 or lower) engaged in their first assignment, two Soldiers at the E4-E5 level who are in their second or later assignment, and a single solider at the E5-6 level (also with more seniority). In a light infantry configuration, they are organized into two fire teams of four Soldiers each plus the squad leader. There are at any time about 7,500 of these squads in the U.S. Army (Active and Reserve, infantry, and other combat arms). This structure largely defines the tactical design of operations large and small in many operational environments, despite the lack of formal study of the effectiveness of this key tactical element since the early 1970s (Melody, 1990).

Alternative squad designs have been adopted even by other units in the U.S. Army. Special Forces employ what is essentially a 12-man squad led by an officer, usually a captain. A warrant officer serves as his second in command, and a mixture of senior and junior non-commissioned officers fill out the unit. While the construction of a basic squad with this level of experience obviously lies outside the realm of feasibility for the larger Army, it is worth asking whether the Army would be better served by squads with a larger fraction of noncommissioned officers (Soldiers above the levels of E3 and E4). It thus seems pertinent to ask whether some alternative manning strategy, in which Soldiers serve longer, have more experience, and operate in tactical units with higher average ranks, could be more effective. If there were compelling evidence of a major increase in effectiveness, then one fundamental way to give dismounted TSUs decisive overmatch would be to upgrade their rank and expertise distributions.

Deficits in TSU and Soldier Training

Lack of time for training before deployment was a common theme in all of the committee's encounters with commissioned and noncommissioned officers. Their observations were replete with a litany of training distractions, including new Soldiers (and leaders) being assigned to a unit late in the training cycle, newly assigned Soldiers lacking essential individual combat skills (e.g., driving combat vehicles), and unavailable training support for new materiel technologies that were issued just before deployment or only after units were in-theater. There were also pleas for new training technologies. But the calls for the new training technologies were not correlated with the training problems recounted. Second only to the need for TSU performance metrics was the need for deliberate, systematic, engineering and management of the TSU training enterprise such that TSUs are able to gain and sustain critical deployment mission skills within the times available. This need for a quantum leap in training effectiveness applies across the skill spectrum from basic rifle marksmanship to language and cultural skills. New technology cannot be used as an excuse for deferring the training of critical skills. The expectation that new technology will be used successfully requires that the Army take full responsibility for the training burden.

Training and Leader Development

The fundamental priority in training is establishing the deliberate and systematic engineering and management that exploits available training technologies and facilities to elevate the TSU's performance to robust deployment readiness levels required of new operational environments. Were such engineering and management of the training enterprise attained, it would then be reasonable to consider, for the longer term, further advances in training technologies

Attaining and maintaining excellent performance by a TSU requires intense, focused training and effective leader development. Yet, as important as training is, for many tasks and missions there are limits to what can be accomplished through a live training exercise.

One general problem is the lack of access to a sufficiently realistic representation of the operating environment. Extreme environments, urban settings, and major cultural differences can be difficult or costly to replicate in a physical training environment, and some operations may simply be too dangerous for live training. Another limitation to conducting live training is the cost and availability of resources such as ranges, training areas, weapons systems, vehicles, fuel, and ammunition. Finally, to make training even more challenging, the range of military operations in the post-Cold War era has expanded to include wide area security as an Army core competency, which adds additional sets of skills to what must be trained, including stability support, cultural competence, interpersonal skills, and some facility with the social networks of the local population.

Next-Generation Simulations and Devices

The live training solutions described above have the advantage of providing the fullest physical experience to the TSU. Trainees use their senses in the most natural way since the environment is mostly real. They are also affected by physical stressors ranging from the exertion while carrying a load and extreme temperatures to the effects of smoke and noise. Nevertheless, the current generation of live training simulations has limitations:

- The lasers used in place of live ammunition cannot shoot through thick smoke, curtains, wooden window shutters, thin walls and doors, etc, which live ammunition would penetrate.
- Wide area weapons (which for training usually transmit a radio frequency signal to indicate personnel within a round's area of effect) do not take into account the protection afforded by urban structures. For example, if the Soldier-trainee is behind a thick wall when a MILES grenade goes off, he may be close enough to receive a strong radio signal, so he is registered as killed or wounded. This is not realistic.
- The ricochets and near misses of real rounds, which would warn a Soldier that someone is firing in his direction, are lacking. So the trainee unknowingly steps out into the "field of fire" and is hit.
- The activity monitoring equipment cannot locate and track personnel because of insufficient GPS signals. This hinders after action reviews of the training exercise.
- The electromagnetic environment to interfere with communications systems is lacking.
- Effects of weapons (e.g., rubble) are lacking.
- Buildings in training exercises are normally constructed of long-term durable materials such as cinder block walls. This limits use of training with future "see through walls" sensors for building materials more likely to be encountered in operational environments.
- Use of steel portable buildings (normally made of steel shipping containers with different facades) may severely inhibit propagation of radio frequency signals, giving unrealistically short transmission distances.
- Buildings on training ranges normally do not have ventilation systems for clearing out smoke, CS gas, natural harmful gases, etc. from the buildings

and underground tunnels after an exercise. This creates problems for repeated training runs.

While it has not been demonstrated that these limitations result in negative training, they make it likely that Soldiers will learn to "game" the training environment, which could result in learning behaviors that succeed in the live training exercise but would fail in actual combat situations.

In addition to issues of realism, fixed physical training sites are expensive to build and populate with live role-players. This limits the amount of training that can be conducted, due to limitations of access and the high cost of training. Finally, there is the issue of availability for a fixed site—it is a resource that has to be scheduled and maintained.

Among the potential advantages of virtual and game-based training systems over live training is the ability of one system to represent a large number of locations, cultures, and scenarios. They are also inherently more accessible than a fixed physical training site since they can be replicated, transported, or used in a classroom or anywhere that a computer can be set up. At the same time, virtual and game-based training systems also have limitations:

- Dismounted transport and action in the virtual world is not natural when it can be achieved just by using a joystick or game controller. Such devices do not require physical exertion, induce fatigue, or permit naturalistic gestures or motions. Solutions such as omnidirectional treadmills and hamster balls are expensive and bulky.
- Interactions with teammates and locals are not natural. For instance, virtual individuals do not look sufficiently different or act individually in ways that enable the trainee to distinguish one from another. Automated characters should perceive, think, act, and react naturally in accordance with a cultural norm, yet still have distinguishable differences.
- There is a general lack of representation of populations of synthetic people that act or react naturally and in accordance with social and cultural norms. Models are lacking of populations of people who interact with one another and belong to complex social networks that influence and are influenced by one another.[12] The simulated population should react to the operations of the TSU in an area, and this should be manifested in the behaviors of groups and individual autonomous characters.
- Current systems do not permit spoken language dialogue with automated characters or natural gestural interaction with other trainees or automated characters.

---

[12]Social networks in the military context are discussed at length in the NRC report on Network Science (NRC, 2005).

- As in live training, simulated urban environments should also present real-world challenges such as electromagnetic interference and occlusions to sensors and communications signals. It should be possible to seamlessly use the same sensors in live and virtual training environments.
- While the potential exists to create any location, culture, or scenario, one of the major barriers to doing so is the general lack of easy-to-use authoring tools to support scenario design, content development, and automated coaching, feedback, and after action review. Current systems that claim to take a data-driven approach tend to only recreate a playback of a scenario as opposed to creating an interactive simulation. To develop content requires more than creating a PowerPoint presentation, which has been the default approach for many years now. Rather, a cognitive task analysis should identify learning objectives, the instruction needs to follow a principled design, and the practical experiences need to be authored for games or simulations. Unfortunately, many of the simulation systems currently available for training require specialized training and a lot of time to develop a new scenario.
- Another barrier to expanding use of this type of training is the cost of component parts, such as the Head Mounted Display in the Dismounted Soldier Training System. High costs drive down the number of systems that can be procured.

Deficits in the Analytical Foundation for Building Decisive TSUs

The lack of an analytical foundation for rifle squad performance limits advances to what is being advocated at the moment by infantry leadership. It precludes a stable architecture for capability development and diminishes the competitive positioning for resources. As discussed in the next chapter, there are no measures of performance or measures of effectiveness for the TSU, no accepted one-sided or force-on-force models, and no widely recognized suite of standard scenarios.

There is no overarching framework to guide the development of Soldier/TSU enhancements. Additionally, there is a lack of U.S. Army Training and Doctrine Command (TRADOC) documentation (e.g., initial capabilities documents or capabilities development documents) for the TSU. It is likely that the same tactics, techniques, and procedures that were effective at squad level in Iraq and Afghanistan may not be optimally decisive against future adversaries. From the committee's view, the TSU should be viewed as a system-of-systems and not merely as a formation. A proper system-of-systems analysis would be able to determine the optimal size (number of Soldiers) and organization (number of fire teams, duties) of the TSU.

The Untapped Human Dimension

Advances in human research and performance have not been fully applied to individual Soldier performance or to small-unit performance. Definitions for human dimension vary greatly among documents and presentations. The prevailing Army human dimension approach is to focus on the cognitive and physical performance of TSUs and Soldiers, however that view is dwarfed by the actual complexities of individual Soldiers and human interactions in teams.[13]

The "human dimension" programs in today's Army consist of underfunded R&D in the Army Research Laboratory, the Army Research Institute, and the U.S. Army Research Institute of Environmental Medicine, plus unfunded, ad hoc, or "interest" activities in TRADOC, the United States Army Forces Command, and the United States Military Academy. Worse, there are no longer any "engineer-equivalent" appliers of science in the TRADOC schools who can understand results of research from the Program 6 agencies and structure programs to implement change. Unlike in lethality, propulsion, or other areas of R&D where Army laboratory results transition to an Army research, development and engineering center, then to co-located program managers, all staffed with engineering professionals, human dimension research lacks a pathway for development and engineering between the researchers and potential end-users.

PROSPECTIVE SOLUTION CATEGORIES

The committee received a great deal of information on the Army's ongoing programs to develop technology options for dismounted Soldiers and TSUs that could potentially contribute to achieving decisive overmatch. Given the capability needs described in this chapter, the committee found that the capability solutions with highest potential to contribute to decisive overmatch for the TSU would likely fall into one or more of five capability improvement areas:

- Designing the TSU
- Focusing on TSU Training
- Integrating the TSU into the Network
- Balancing TSU Maneuverability, Military Effects, and Survivability
- Leveraging Technology Advances in Portable Power

In the following chapter, the committee presents overarching recommendations on what will be needed to realize the potential of capability solutions in these areas or any others. Chapter 4 returns to these five areas to explore options in each area that the committee judged to have the most promise.

---

[13]Many of these complexities are detailed in the NRC report on Neuroscience Opportunities for Future Army Applications (NRC, 2009).

REFERENCES

AUSA (Association of the United States Army). 2011. The U.S. Army Squad: Foundation of the Decisive Force. Available online www.ausa.org/ publications/ ilw/Documents/TB_The_Squad_web.pdf. Accessed March 12, 2013.

Brown, MAJ R.B. 2011. The Infantry Squad: Decisive Force Now and in the Future. Available online http://usacac.army.mil/CAC2/MilitaryReview/ Archives/English/MilitaryReview_20111231_art004.pdf. Accessed March 29, 2013.

CALL (Center for Army Lessons Learned). 2003. The Modern Warrior's Combat Load, Dismounted Operations in Afghanistan April - May 2003, Task Force Devil, Coalition Task Force 82, Coalition Joint Task Force 180, OPERATION ENDURING FREEDOM III. Available online http://thedonovan.com/archives/modernwarriorload/ModernWarriorsCombatLoadReport.pdf. Accessed February 13, 2013.

Chait, R., A. Sciarretta, and D. Shorts. 2006. Army Science and Technology Analysis for Stability Operations. Washington, D.C.: National Defense University, Center for Technology and National Security Policy October.

Chait, R. A. Sciarretta, J. Lyons, C. Barry, D. Shorts, and D. Long. 2007. A Further Look at Technologies and Capabilities for Stabilization and Reconstruction Operations. Washington, D.C.: National Defense University, Center for Technology and National Security Policy.

Endsley, M., S. Selcon, T.D. Hardiman, and D. Croft. 1998. A comparative analysis of SAGAT and SART for evaluations of situation awareness. Presentation at the 42nd Annual Meeting of the Human Factors and Ergonomics Society. Chicago, Ill.: Annual Meeting of the Human Factors and Ergonomics Society.

Finn, P. 2011. A Possible Future for Drones: Automated Killings. The Washington Post. Available online www.washingtonpost.com/national/ national-security/a-future-for-drones-automated-killing/2011/09/15/gIQAVy9mgK_story.html. Accessed March 12, 2013.

HQDA (Headquarters Department of the Army). 2007. Field Manual FM 3-21.8. The Infantry Rifle Platoon and Squad. Seattle, Wash.: CreateSpace Independent Publishing Platform.

Matthews, W. 2008. Helping info flow freely: Insurgents outdo U.S. military, says U.S. 3-Star. Defense News 23(4): 16.

Melody, MAJ P.E. 1990. AD-A225-438. The Infantry Rifle Squad: Size Is Not the Only Problem. Fort Leavenworth, Kans.: School of Advanced Military Studies United States Army Command and General Staff College.

Miller, N.L., L.G. Shattuck, and P. Matsangas. 2011. Sleep and fatigue issues in continuous operations: A survey of US Army officers. Behavioral Sleep Medicine 9: 53-65.

NRC (National Research Council). 1997. Energy-Efficient Technologies for the Dismounted Soldier. Washington, D.C.: National Academy Press.

NRC. 2004. Meeting the Energy Needs of Future Warriors. Washington, D.C.: The National Academies Press.

NRC. 2005. Network Science. Washington, D.C.: The National Academies Press.

NRC. 2009. Opportunities in Neuroscience for Future Army Applications. Washington, D.C.: The National Academies Press.

U.S. Army. 1992. Field Manual 7-8. Infantry Rifle Platoon and Squad. Washington, D.C.: Headquarters, Department of the U.S. Army.

U.S. Army. 2011. ADP 3-0. Unified Land Operations. Arlington, Va.: Headquarters, Department of the U.S. Army. Available online http://armypubs.army.mil/doctrine/DR_pubs/dr_a/pdf/adp3_0.pdf/. Accessed March 12, 2013.

U.S. Army. 2012. ADRP 3-0. Unified Land Operations. Available online http://armypubs.army.mil/doctrine/DR_pubs/dr_a/pdf/adrp3_0.pdf. Accessed March 12, 2013.

Zoroya, G. 2011. Injuries Cost More Troops Their Limbs. USA Today. Available online www.usatoday.com/news/military/story/2011-09-19/troops-injuries-war-casualties-amputations-lost-limbs/50472074/1.

3

Setting the Conditions to Achieve Soldier and TSU Overmatch

Prospective solutions to meeting the capability needs described in Chapter 2 would have various potentials to contribute to the goal of achieving tactical small unit (TSU) overmatch. However, no principled means exists to evaluate all solution candidates and implement the ones that would contribute the most. It would be far too tedious a process to evaluate all conceivable solutions against the Doctrine, Organization, Training, Materiel, Leadership and Education, Personnel and Facilities (DOTMLPF) domains and even less practical to implement all of them via the materiel acquisition process.

This chapter discusses four essential actions needed to set the conditions for the Army to exploit the potential available for TSU overmatch. The following conclusions form the basis for these actions:

- In most areas of potential improvement in the human dimension sphere of TSU performance, the Army's modest funding of human performance research and development (R&D) in relatively narrowly delimited domains has severely limited TSU improvement options. The Army will need to significantly increase investments in human dimension research, development, and engineering to provide a more robust menu of decision options. This required level of emphasis is discussed in the first section of the chapter.
- The Army should have an analytically sound approach for evaluating combinations of potential capability options holistically, rather than evaluating options independent of each other (as stovepiped "eaches") without considering the TSU and Soldier as functional wholes. The second section of this chapter argues that the established disciplines of systems engineering and system-of-systems engineering are applicable for such evaluations—*provided* the enriched and comprehensive concept of the human dimension (as discussed in the first section) is fully incorporated into the systems engineering methodology.
- To support fully the anticipated benefits of both a richer palette of potential solutions and an analytical systems engineering approach, the Army must employ a rigorous methodology for developing a comprehensive set of measures of effectiveness (MOEs) and measures of performance (MOPs) that better represent the mission performance of a TSU, including the capabilities and limitations of all relevant components—the individual Soldier, materiel, human dimension, doctrine, and organization— and their interactions, in objective terms. The third section of the chapter discusses this need for more objective measures that are

directed not just toward a single component, such as the human dimension, but at the entirety of the TSU ensemble, which includes all component dimensions and their interactions.

- The committee doubts that solutions to achieve overmatch capabilities can be successfully implemented with the Army's typical acquisition approaches because the principled groundwork necessary to analyze the TSU system has not been laid for a natural progression from design to experiment/development and then to acquisition, test, and fielding. In particular, the usual solution space of the DOTMLPF domains has traditionally constrained the available options and programmatic implementation to a predominantly materiel acquisition process. Accordingly, the committee urges the Army to tailor its acquisition processes—within the legally mandated acquisition system—to ensure that satisfactory solutions are developed and fielded rapidly, with a full complement of training and support. The last section of this chapter describes what the committee views as obstacles and weaknesses embedded in recent/current acquisition practices and processes and what it suggests as ways to overcome them.

PLACING EMPHASIS ON THE HUMAN DIMENSION

The committee's expectation that the greatest returns on TSU investments will come from more thorough integration of the human dimension with materiel advances was discussed in Chapter 1. This expectation derives from the statement of task (including the clarifying guidance from the sponsor) combined with the committee's awareness, illustrated throughout Chapter 2, of the new emphasis being placed on both the tactical and strategic importance of the dismounted TSU and Soldier in current and expected future Army missions. Since Soldiers touch everything the Army is and does, a challenge is to determine how the whole of today's panoply of human dimension programs might be recast to give new emphases that will lead to dismounted TSUs and Soldiers with decisive overmatch.

As discussed in chapter 1, the committee arrived at the following working definition for the human dimension:

As used in this study, the human dimension means all of the attributes of the individual Soldier and of the collected Soldiers of the TSU that impact performance of mission tasks. These include the skills, abilities, and knowledge brought with them into the Army upon recruitment, even from prior education or job experiences; personality traits; individual and collective military training; skills, abilities, and knowledge from prior military assignments; TSU command chain leadership; unit social environment including morale, cohesion, and emotional state; the ergonomic design or human factors engineering of the Soldier-machine interfaces; as well as locale acclamation (time zone, elevation, temperature, etc). Skills, abilities, and knowledge include the physical, mental, and emotional. Bearing with real impact but less directly on mission task performance are the domestic or family environments of each Soldier, which have not been included here. Nor has the committee included issues of morality that may bear on overall mission accomplishment in a strategic sense but not on tactical tasks, except as morality issues may influence the effectiveness of the unit leadership chain or the health of the unit social environment.

During its information gathering, the committee heard the human dimension referenced and used in a variety of ways that fit somewhere in this definition but seldom if ever covered it comprehensively.

There are R&D programs in the Army Research Institute for the Behavioral and Social Sciences and in the Human Research and Engineering Directorate of the Army Research Laboratory on individual and collective training, leadership, and personnel, but none are adequate to the TSU overmatch challenges. Even if the individual DOTMLPF domains most relevant to the human dimension (e.g., doctrine, organization, training, leadership, and personnel) were adequately addressed in research and analysis, there are questions with exciting potential left hanging. For example, if selection instruments could make it more likely that accessing Soldiers have complementary temperaments, do the TSU leadership challenges in today's theaters fade, or do they transform into more complex issues? If doctrine provided for more robust on-call fires and logistics support to dismounted TSUs, how should the organization structure exploit the opportunity for load-carrying ability to be a less-critical factor, especially for first-term Soldiers? Would such doctrine and organization changes allow a change in the personnel, possibly to longer-serving, more skilled Soldiers in the TSU? Unfortunately, the committee could find no research or analysis programs addressing any such interactions among the DOTMLPF domains.

The Army G1 (Deputy Chief of Staff for Personnel) and the Assistant Secretary of the Army for Acquisition, Logistics, and Technology share responsibility for the MANpower, PeRsonnel, INTegration (MANPRINT) program, which seeks to ensure that key domains of Soldier-related issues are considered in the design of materiel systems. While integration across the domains is quite broad and doctrinally begins prior to formal materiel program inception, MANPRINT responds primarily to materiel program needs. MANPRINT as constituted currently would not be expected to seek an optimum TSU configuration, be concerned with TSU collective training, or be concerned with social dynamics within a TSU. MANPRINT's influence is on the course of Army *materiel acquisitions*, although some of the analytical tools produced for the MANPRINT program could be applied to TSU design and evaluation that fully incorporates the human dimension.[1]

The Army Medical Corps, led by the (Army) Surgeon General, has responsibility for Soldier (and Soldier-family) physical (and increasingly, mental) health and especially the restoration of performance following injury. With R&D interests that are often colloquially delineated as "skin-in", the (Army) Surgeon General's interests focus on

---

[1]The MANPRINT domains include manpower (the number of Soldiers required), personnel capabilities (Soldier cognitive and physical capabilities), training; human factors engineering (design of Soldier-machine interfaces to reduce errors, improve performance, and reduce cognitive or other selection demands), system safety (reduce human and machine contributions to accidents), health hazards (chronic risks such as those regulated in civilian occupations by the Occupational Safety and Health Administration), and Soldier survivability (minimizing injury either from the environment or when a platform or unit engages in combat) (U.S. Army, 2001; U.S. Army, 2012).

avoiding performance debilitation rather than maximizing unit performance. Products of Army Medical Department R&D are fielded primarily through the medical practice of Army Medical Department personnel (physicians, nurses, etc). Some medical R&D results influence the design of rations and other programs of the Natick Research, Development and Engineering Center.

In sum, the Army lacks a working definition for the human dimension that is clear and comprehensive enough to inform and guide all of DOTMLPF, five elements of which involve human considerations. As a consequence, there is no single proponent for the TSU, and the Army lacks an in-place, practiced tool of implementation—that is, an administrative organization structured to implement new human dimension-centered technologies into Army practice. This limitation is applicable to all Army acquisition programs, but the scope of this report is limited to the TSU.

Finding: An essential principle for achieving overmatch capabilities is to recognize that the human dimension is at the core of all dismounted Soldier and TSU improvements.

Finding: Existing Army R&D programs are insufficiently resourced to provide a range of human dimension technology opportunities that could be selected to provide overmatching TSU performance.

Finding: The current niche organization of research, development, and engineering tends to preclude exploration of interaction opportunities among the human dimension-related domains of DOTMLPF.

Finding: The Army lacks an engineering-like function to orchestrate the transition of results of human dimension research into operational requirements.

Recommendation 1: To determine overmatch options for the tactical small unit, the Army should provide sufficient resources for the full range of human-dimension opportunities and solutions that might provide overmatching performance.

SYSTEMS ENGINEERING FOR DECISIVE OVERMATCH

The Statement of Task discusses weapons with overmatch capability—the M1A2 tank, the F-22 fighter, and the Seawolf attack submarine—as a point of departure for considering overmatch capability for the dismounted Soldier and TSU. Such weapon systems consist of a number of complex subsystems that interact and can be considered to be interdependent. A commonality of all three weapon systems is that they employed systems engineering methodologies during conceptualization and development to determine the configurations of the various subsystems that were best-suited to meet a top-level set of performance and effectiveness metrics for the system as a whole. Since the Army considers the Soldier to be a system (and the committee concurs) and multiple individual Soldier systems constitute a TSU, it is reasonable, by extension from the three referenced overmatch systems, to assume there is value in using systems engineering methodologies when considering improved capabilities for both the Soldier and the TSU. Perhaps in recognition of the above situation, the Statement of Task explicitly directs the committee to examine the applicability of systems engineering to Soldiers and small units. This section addresses that topic.

As currently structured, the Army's dismounted TSU is a squad consisting of nine Soldiers organized into two four-Soldier fire teams, led by a squad leader. The capabilities of the TSU are thus a function of the capabilities of its members, how they are organized, their equipment, their training, and their leadership. The individual Soldiers may also be viewed as complex systems. Their basic attributes—physical, psychological, cognitive, sensory and perceptual—differ. Their individual materiel—clothing, weapons, body armor, sensors, communications devices, rations, etc.—impose upon them physical, psychological, cognitive, sensory, and perceptual loads that interact with training and leadership in determining their capabilities. Dependencies are numerous and complex, and it is important that the magnitude of the loads be managed in a top-down, holistic fashion to ensure that a balance is maintained among technologies that may enhance one aspect of capability at the expense of another.

The TSU is an even more complex system than the Soldiers in it. It is composed of individuals who must perform demanding collective tasks, including interfaces with supporting capabilities external to the TSU. It is equipped with materiel that performs unit-level functions rather than an individual function and that may require the cooperative actions of two or more Soldiers for optimal performance. The degree to which individuals are integrated into the small social element that is the TSU is important to collective capabilities. The organization chosen for the TSU provides a framework for decomposing collective tasks into components performed by individual Soldiers and for assigning different loads—again physical, psychological, cognitive, sensory and perceptual—to the members of the unit. Collective training establishes not just collective capability but also the bonds between Soldiers and between Soldiers and leaders. Dependencies that develop as a consequence of doctrine, organization, leadership, and training are critical determinants of decisive overmatch.

If a balanced approach is to be taken to identifying capability options that will make the TSU decisive, then the components of the TSU—the Soldiers—should be treated as systems, and the TSU should be treated as a system of systems. A holistic, top-down perspective should be used, dependencies should be identified and accounted for, and attention should be focused on both the enhanced performance offered by an option along one or more dimensions and the degradations along other dimensions caused by its introduction. Finally, the issue of budgets cannot be ignored. A means of trading between alternative integrated sets of options must be developed that facilitates answering the question "Given additional funds, on what should they be spent?" and the question "Given two alternative option-sets with different capabilities, which is preferred?"

Unlike developments of the Abrams tank or Seawolf submarine, which were a consequence of latent technology advantages and proven science and engineering including a well-developed system analysis, system engineering, and system-of-system analytic capability, the Army should not expect that developing human systems will be anything but challenging. It will not have a rigorous body of physics and engineering principles to apply to the task. But the development of such methodology has long been overdue and can be continually applied to ensure overmatch capabilities for Soldier systems well into the future.

Previous Army studies have advocated that a systems engineering methodology is important, given that the Army views the Soldier as a system. Appendix D discusses these studies in that context; the appendix also presents indicators that a systems engineering approach has thus far not been effectively implemented, despite the conclusions of these studies. The committee believes that the Soldier and the TSU must be treated as a system and a system of systems, respectively, if improvements in capability that synergistically provide overmatch are to be achieved. As indicated in the first section of this chapter, the committee attaches equal or greater importance to the integration of an expanded concept of the human dimension into the systems engineering approach to the Soldier and the TSU.

The goal of integrating the human dimension with a systems engineering approach is similar to that of *human-systems integration,* which builds upon and expands preceding work in human factors research, ergonomics, cognitive engineering, and other disciplines in order to focus on how human beings perform tasks using modern technologies and complex systems. Unfortunately, the programs in the Department of Defense that incorporate human-systems integration have fallen short because the principles of integration are applied too late in the development process and overlook aspects of the human dimension that are critical for TSU performance. See Box 3-1.

Box 3-1 Military Implementation of Human-Systems Integration

The military services have previously addressed limited elements of the human dimension in a systems context under the broad nomenclature of "human-systems integration" (HSI). Currently, the Air Force lead for HSI efforts is the Air Force Human Systems Integration Office, and the Navy lead program is at Naval Surface Warfare Center (NSWC) Dahlgren.[a] The Army continues to fulfill its HSI requirements through the MANPRINT program, which dates to 1986 (U.S. Army, 2009).

The HSI concept was intended to capture the full scope of work associated with accommodating people in systems, and much work published under this rubric has been concerned with integrating human-systems design considerations into the larger domain of systems engineering (for example, see Booher, 2003). However, the Army MANPRINT program focuses on the materiel acquisition process—the process that begins after requirements and specifications have been set for the materiel subsystems of what should be an integrated, human-based and materiel-equipped system. Unless and until the Army rigorously applies HSI principles across the full scope of task analysis (including, for example, communication protocols, mission planning, and concepts of operations for all potential tasks across the range of military operations), requirements definition and specification grounded in these task analyses, system design to those requirements, and acquisition of the integrated system—rather than just applying HSI to isolated materiel acquisition programs—MANPRINT will continue to achieve little more than adjusting the interfaces between the human and the tools of the Soldier's trade. Furthermore, the committee believes the human dimension, as defined and discussed in this report, covers some nonmaterial elements of the DOTMLPF domains, such as leader development and small group dynamics, that are typically not addressed in the HSI context but are critical to overmatch for dismounted TSUs.

[a] See the Air Force Human Systems Integration Office website at ww3.safaq.hq.af.mil/organizations/afhsio/index.asp. HSI at NSWC Dahlgren is described at: www.navsea.navy.mil/nswc/dahlgren/ET/HSI/default.aspx.

To identify technologies that would make the Soldier and the TSU decisive on future battlefields, a holistic perspective must be applied. Desired capabilities of the Soldier and the TSU must be described in terms that permit progressively more detailed functional and task decomposition, followed by the assignment of solutions to collective and individual tasks to meet associated task requirements. It is worth noting that any solutions to the challenge of designing the TSU can be viewed in terms of one or more of the DOTMLPF domains: doctrine, organization, training, materiel, leadership, personnel, and facilities. Also, the objective function of the solutions must include robustness, versatility, resilience, and agility, to guard against the fragility of a single-point solution.

From the perspective of systems engineering, capabilities desired must be described in terms that can be quantified. Relevant examples are the time required to complete a mission, the residual capacity to undertake follow-on missions, or Soldier and TSU agility and versatility. These capabilities must be described in the context of one or more scenarios that incorporate the Army's standard planning parameters of METT-TC (Mission, Enemy, Terrain and weather, Troops and support available—Time available, and Civilians). In addition, the capabilities must be jointly feasible as an integrated DOTMLPF solution.

Finding: The Army has consistently described the Soldier as a system (implying the TSU is a system of systems), and previous studies have concluded that the Army should use a systems engineering methodology for the Soldier (see Appendix D). Nevertheless, the committee found no evidence that these conclusions had been acted upon in a comprehensive manner. Moreover, these previous studies were framed largely in the context of providing enhanced capability via materiel solutions, whereas the committee has observed that overmatch capability can best be achieved by considering the full spectrum of applicable DOTMLPF domains, making even more important the need for a full-spectrum systems engineering capability in support of the Soldier/TSU.

Qualified system engineering professionals, possibly with centralized leadership located in the Office of the Assistant Secretary of the Army for Acquisition, Logistics and Technology and distributed among the Army Research, Development, and Engineering Centers, would be well positioned to trade off candidate solutions within or among the various DOTMLPF domains while ensuring that specified and required capabilities are achieved. Part of the staff for this function could also be located at the Maneuver Center of Excellence to ensure that an integrated view of the Soldier and the TSU is considered for all requirements, functions, architectures, and designs.

Recommendation 2: The Army should establish a Systems Engineering executive authority to support a system-of-systems engineering environment that will be responsible for developing methodologies and analytical tools to evaluate and acquire total system solutions for the dismounted Soldier and TSU. This executive authority must have sufficient seniority, influence, and budget control to operate effectively across the entire Army acquisition community (including research and development, test, and evaluation) in executing its systems engineering mission.

METRICS FOR THE DISMOUNTED TSU AND SOLDIER

The decomposition of high level requirements and functions into concrete, measurable system attributes depends upon numerical values or metrics, which allow the systems designer to create and evaluate alternative solutions. At the lowest level of capability analysis, options for improving a capability can be characterized by attribute or parameter values. For the Soldier, these include, for example, performance measures of strength; endurance and load carrying ability; personality dimensions; cognitive, sensory and perceptual abilities; and auditory performance, as well as the standard measures of height, weight, etc. These parameters or attributes, combined with metrics linked to the other dimensions of DOTMLPF, determine how well a given task will be performed under given conditions: how long it will take, how accurate and complete the result will be, how much energy will be consumed, etc. These are MOPs, which quantify how well the Soldier or a TSU performs a task or sequence of tasks. An example would be engaging an enemy with an individual rifle or combined fires of a TSU at *x* meters and achieving a kill *y* percent of the time. Beyond the assessment of task performance for the Soldier and TSU are MOEs, which assess changes in behavior, capability, or operational environment. MOPs measure what is accomplished and help to verify whether objectives, goals and end states are being met—for example, achieving kills *x* percent of the time at *y* meters decreases TSU vulnerability to enemy small arms fires by z percent.

Associated with each MOP and MOE are acceptability criteria, which set threshold (minimal acceptable) and objective (desired) levels. It is also important to understand that MOPs and MOEs are not only associated with individual Soldiers: TSUs also have MOPs and MOEs. A focused discussion of metrics is contained in Appendix E.

Metrics play an essential role in defining what makes the dismounted Soldier and the TSU decisive and in selecting and evaluating combinations of technologies that would constitute a responsive solution. However, just listing a set of metrics is not enough: the real issue is how metrics are developed, defined, and used in the requirements definition, experimental analysis, acquisition, and test/evaluation processes. As explained with examples in Appendix E, the Army needs metrics that can be applied not just to one item of equipment or materiel solution; the same metrics should be appropriate for use, and applied in practice, across capability options that draw on different combinations of DOTMLPF domains and different approaches within those domains. That is the only way to provide an objective basis for comparing different combinations of possibilities to find the most satisfactory approach for decisive overmatch and to ensure that the approach includes robustness, versatility, resiliency, and agility—attributes necessary to guard against single-point, fragile solutions.

Both the Soldier and the TSU are complex systems of systems composed, at any given time, of humans and materiel and their extended network. The Soldiers and TSU (and someday, learning, autonomous systems) are trained to accomplish their missions in a particular organization according to established doctrine.[2] The degree to which a particular mission is accomplished is further affected by such factors as local acclimatization, degree of sleep loss, leadership, social comfort in the small unit, and other

[2]Doctrine is normally considered to be aligned with operations at levels above the TSU. At the TSU level, doctrine is embedded in operational tactics, techniques, and procedures.

critical elements of the human dimension. If technologies are to be evaluated and design trades made, methods must be available that address the interdependencies among the DOTMLPF attributes of proposed solutions and task and mission performance. The Army does have a tried and tested methodology and a mature set of metrics for Armored Systems and Mounted Combat that, together with models and simulations, can predict or estimate engagement, battle, and campaign outcomes for a given set of performance data and conditions. However, a similar methodology and proven metrics do not exist for the Soldier or especially for the TSU, in dismounted operations such as maneuver warfare or wide area security.

Quantitative models for cost and performance interactions are routine tools of design and engineering for all components of the TSU—except for Soldiers! Such models of anthropometry (body shape and size) have been used for decades in design of military vehicle crew stations, individual weapons, and individual protective equipment. But, the extension of these models to be useful tools in the systems engineering of TSU ensembles has been limited by funding.

In all military services, models predicting crew task performance, including cognitive workload, as functions of operator or crew station design have been key design tools for military vehicles from self-propelled howitzers to aircraft. However, the committee could find no evidence that these models have been considered in designing information technology systems and networks for TSUs.

The interdependencies among the human and materiel aspects of solutions and the various Soldier and TSU capabilities are numerous, complex, and very important. The Army must recognize that any change that improves some aspect of performance or effectiveness will almost certainly impact others because, as stated earlier, Soldiers and TSUs are integrated systems of systems. Significant design trades must be made in the realm of Soldier and TSU, in part because missions have grown more complex, but also because of the potential gains associated with integrating the Soldier and TSU into the Army network. For example, it is now possible, and capabilities will continually grow, to develop Level 1 situational awareness via both organic assets and feeds from adjacent and higher echelons. but this increase in information input comes at the cost of a higher cognitive load. Similarly, the operation of unmanned air or ground vehicles in the TSU may require that organization be changed to ensure that the addition of cognitive and physical tasks does not degrade performance in some other area. As a third example, the potential need to carry additional gear incorporating new technologies may compromise performance over longer duration missions. Options to increase survivability via body armor may appear attractive but if weight is increased, tactical mobility will be compromised and incidences of skeletal-muscular injuries may increase. As a final example, sharing enhanced situational awareness through materiel and human dimension capabilities improves situational understanding (and therefore decision-making) for the TSU as a unit, as well as for the individual Soldiers.

The committee was made aware of ongoing Army efforts to develop MOPs and MOEs for the TSU but was unable to gain any insights into efforts underway. The committee can thus only emphasize how critically necessary a revolutionary metrics development approach is for supporting a rigorous assessment of the integration of all components of the TSU: individual Soldier, materiel, human dimension, doctrinal, and organizational—and their interactions. Clearly, some existing MOPs such as probability of kill at range or sprint speed across a gap will remain germane. But performance factors such as degraded TSU maneuverability, incidence of Soldier injury caused by loads, or degradation of situational understanding due to fatigue are even more important for assessing decisive action. New factors such as face-to-face communications skills with civilians and conflict resolution skills are now part of security tasks for missions both present and future. From a human dimension perspective, MOPs and MOEs for the TSU must address the impact of organization, leadership, training, and personnel on small unit performance in both the short term and long term. Assessments of new capabilities—for instance, the integration of the TSU into the Army information network—cannot merely measure the performance of a single enabler, such as the materiel interface to an information system, but must rather evaluate all the accompanying doctrinal use, organizational assignment, training, leadership abilities, and personnel skills that must be considered in developing MOPs and MOEs aimed at ensuring TSU overmatch.

The committee recognizes that a rigorous methodology will not happen overnight for developing and maturing MOPs and MOEs that: (1) address the integration of all aspects of Soldier and TSU enhancements, plus their complex interdependencies, and (2) enable objective, validated predictions of Soldier and TSU performance and effectiveness. However the committee was made aware of a significant body of research that has explored the relationships among attributes of Soldiers and TSUs and performance. This research, primarily in the area of TSU capabilities, should be assembled and brought to bear, recognizing that in many cases its application will produce only marginal improvements until larger investigations can be undertaken that address multiple variables simultaneously, in the field or in the lab, to better understand the TSU as an integrated system of systems.

In support of an analytical systems engineering approach, the Army must develop a rigorous methodology as well as a comprehensive set of MOEs and MOPs against which to measure performance and degree of mission success. The Army Warfighting Experiments and the Combat Training Centers may provide venues for opportunistic data collection, particularly when considering topics for which some narrow research results already exist but have not been integrated into the Army's knowledge base. However, there will be topics for which rigorously designed and executed experiments will be necessary, and those might compromise the objectives of a particular Army Warfighting Experment or the Combat Training Centers. Perhaps more important for the subject of this report, the lack of MOPs and MOEs that realistically assess both human and materiel contributions to required capabilities has vitiated real progress toward holistic design and evaluation of the TSU and Soldier, despite a history of advice to that end (see Appendix D).

With regard to the development of a methodology and accompanying metrics, the Army must ensure that it reviews the appropriateness of traditional metrics as well as developing new and innovative metrics development processes that adequately relate to the envisioned Soldier and TSU system-of-systems concepts. Some traditional metrics such as "loss exchange ratio" may not be the most appropriate for assessing the impact of various human dimension characteristics—for example, improved selection of personnel, better training, better leadership—whereas other traditional metrics previously used to assess materiel systems (e.g., impact on decision times, used to evaluate mission command systems may also be adequate for analysis of non-materiel enhancements. New metrics—for example, ability of squad members to adapt rapidly to the sudden loss of direct and distant leadership with no loss of momentum or decisive advantage—will likely need to be developed. Above all, the Soldier as "end user" should be included at both ends of the design process by contributing to the formulation of MOPs and MOEs on which the design is based and by participating in operational testing of concepts and prototypes during subsequent development and acquisition phases. The description of such a methodology, as well as the role of appropriate MOPs, MOEs, and indicators, is important enough to warrant a fuller discussion, which is provided in Appendix E.

Finding: A rigorous systems engineering methodology and an accompanying comprehensive set of measures that better represent the performance and effectiveness of a TSU are required to fully support the anticipated benefits of both a richer palette of potential solutions. This would include the capabilities and limitations of all of the components—materiel, human, and other dimensions—and their interactions, in objective terms. This need for more objective measures is directed at the entirety of the TSU ensemble that includes human and materiel dimensions, as well as other dimensions and the interactions among them.

Recommendation 3: The Army should develop, maintain, and evolve an optimal set of measures of performance (MOPs) and measures of effectiveness (MOEs) for assessing capability improvements for the dismounted Soldier and TSU by investing in an analysis architecture and infrastructure, including a comprehensive metrics development methodology that supports objective dialogue among combat and system developers, systems engineers, trainers, and program activities. The MOPs and MOEs, together with the guidance for using them, should be tested and validated for practical application and ease of use, as well as for accuracy as predictors and indicators of desired performance and effectiveness outcomes.

STREAMLINING ACQUISITION OF SOLUTIONS TO ACHIEVE TSU OVERMATCH CAPABILITIES

As noted in the previous two sections and discussed more fully in Appendix D, multiple studies have advised the Army to train, equip, and sustain the dismounted Soldier as a holistic entity or system, rather than as a user of independent materiel components or "piece-parts." Yet the committee found limited, if any, evidence that the concept has been implemented within the Army. As noted above, the committee was made aware of an effort to develop MOPs and MOEs appropriate for the range of operations expected of a dismounted TSU, but it was unable to determine the nature of the effort or if it was still active. There is no evidence of the artifacts one would expect to find if a systems engineering approach were being executed. For example, the committee found no architectures for the Soldier as an integrated whole, nor for the TSU. Requirements were found to be incomplete and design criteria not yet developed. Artifacts such as weight tapes may exist for specific materiel systems but do not exist for the full sets of Soldier or TSU equipment. Functional decomposition and task analyses beyond the level of the Army universal task list do not appear to be in use by the various agents responsible for training, equipping, or sustaining the Soldier and the TSU. There does not appear to be a single authority for defining trades among the DOTMLPF attributes of the Soldier or the TSU and deciding on solutions that are consistent with a holistic, integrated approach. Given this state of affairs, the committee concluded that there are significant barriers to treating the Soldier and the TSU as systems or systems of systems. The committee believes that one of these barriers is the current set of acquisition practices, as discussed in the remainder of this chapter.

For more than two decades, the Army has used the term "Soldier as a System" to describe a holistic approach to developing, procuring, and supporting Soldier capabilities. Yet the committee found no evidence of Army processes that routinely and systematically consider the interdependencies and synchronization between the squad and (small) unit capabilities, and no program of record exists that supports the approach envisioned either in reports released in 1991, 2000, and 2006 (see Appendix D) or in the deliberations of this committee. Instead—at least as of the fiscal year 2010 budget—development, procurement, and support for the dismounted Soldier and the TSU were defined through more than 70 programs of record. During the past 10 years, rapid fielding to deployed or deploying TSUs and Soldiers—typically accomplished outside the formal acquisition process in order to meet urgent wartime capability gaps—has exacerbated the historical piecemeal approach to outfitting Soldiers for their roles in the tasks and missions of a dismounted TSU.

One symptom of the problem lies in the requirements process. The committee was unable to identify in existing Army requirements generation and acquisition processes an integrated assessment methodology (and associated tools) adequate for evaluating desired enhancements to the physical and cognitive performance and mission effectiveness of either the individual dismounted Soldier or a dismounted TSU. A second symptom of the problem is associated with the acquisition system—or at least with current practices within that system. The committee found little evidence of the ties that should exist across the design, development, and acquisition of materiel for the dismounted Soldier and TSU.

Progress is unlikely to happen unless two fundamental changes are made, associated with the two symptoms described above. The Army must create a single, formal, system-of-systems program of record at the TSU level with appropriate authority and budget. Second, consistent with the recommendations of the Final Report of the 2010 Army Acquisition Review, a "collaborative requirements process" should be established under the leadership of the U.S. Army Training and Doctrine Command (TRADOC) to develop requirements in a holistic, integrated fashion for the TSU and the Soldiers in it. Only if single, overarching leadership is formally established with sufficient authority and budget is it likely that the necessary systems approach will finally be implemented. That leadership should review Army acquisition processes to determine if they should be changed. For example, does the equipping process as embodied in the Army Force Generation paradigm mean that it is no longer appropriate to think of materiel buys for the total force? Should designs no longer accommodate the "lowest common denominator" of Soldier capability? Instead, does creating a dismounted TSU and Soldier with decisive overmatch capabilities require assigning higher quality recruits to the TSUs and designing materiel with their higher quality in mind?

Compounding the shortcomings of the requirements generation and acquisition processes, the committee believes that the MOPs and MOEs that do exist are much too simplistic (e.g., MOP of basic rifle marksmanship, MOE of loss-exchange ratio) to assess the complexities (especially cognitive aspects) of an integrated Soldier/TSU analysis effort. Ties that should exist across design, development, and acquisition of materiel systems for the dismounted TSU and Soldier just do not exist.

An integrated acquisition approach for all Soldier/TSU systems is lacking. For example, the portfolio of the Program Executive Office-Soldier (PEO-Soldier) includes lethality (personal weapons), survivability (personal armor), and operating systems (clothing, parachutes) but not direct control of important areas such as human dimension considerations, communications systems, sensor systems, and robotic systems. Additionally, Product Manager, Ground Soldier is located in Fort Lewis, Washington, rather than near the TRADOC subject matter experts at the Maneuver Center of Excellence in Fort Benning, Georgia.

The acquisition strategy implicit in presentations to the committee from PEO-Soldier and the Maneuver Center of Excellence is built on whole-of-Army buys. This strategy is too ponderous and slow for rapidly advancing communications and information collection and networking technologies.[3] It leads to buying capability solutions that are either inadequate for the range of perceived threats or too expensive and lengthy to be affordable and practical investments.

The array of technologies available to the Army constitutes an impressive set of potential opportunities to improve the capabilities of the dismounted Soldier and TSU. However no one technology solution in isolation is capable of achieving consistent overmatch, and for each technology solution there is a danger of unanticipated consequences of varying degree unless a holistic approach is taken to evaluating and selecting innovations and improvements.

In the committee's judgment, a key action that the Army can take to facilitate improving capability and achieving overmatch is to focus on the acquisition process. Responsibility and authority for Soldier and TSU research and development must be centralized. The committee is not the first body to make this recommendation, but its importance—in this period of the central role of dismounted infantry and constrained budgets— is if anything greater than at any time in the last two decades.

Finding. Despite multiple advisory reports, extending back more than two decades, on the critical importance of a holistic approach to developing, procuring, and supporting Soldier capabilities, the Army is still acquiring kit and gear for the dismounted Soldier through separate programs of record (70 separate programs in the fiscal year 2010 budget). Army acquisition essentially consists of providing for independent efforts to support the TSU and Soldier, rather than providing for integrated systems. The urgency to support the force in the field during current operations has led to a reliance on rapid equipment fielding, which has exacerbated this stove-piped approach.

It is questionable that solutions to achieve overmatch capabilities can be successfully implemented with the Army's typical acquisition approaches because the principled groundwork for a natural progression from analyzing the TSU as a system has

[3]The "communications and information collection and networking technologies" to which the committee is referring includes all those previously included under the military rubric (now replaced) of C4ISR (command, control, communications, and computing; intelligence, surveillance, and reconnaissance).

not been laid. In particular, the usual solution space of DOTMLPF presumes available solution options and programmatic implementation via a predominantly materiel acquisition process.

The approach of acquiring and fielding every "new" technology to the entire Army has become both impractical and unaffordable. It runs counter to processes tailored to the need for more "rapid fielding" in Iraq and Afghanistan, and is especially counter to fielding in support of dismounted TSU deployments.

Finding: The Army acquisition processes can be tailored—within the legally mandated acquisition system—to develop and field solutions optimized for system-level effectiveness with a full complement of training and support.

Recommendation 4: The Army should establish an executive authority for TSU integration, responsible for option generation and evaluation, requirements currency, and programmatic acquisition for the Soldier and TSU within a metrics-driven, system-of-systems engineering environment.

REFERENCES

Booher, H.T. 2003. Handbook of Human Systems Integration, New York, N.Y.: John Wiley and Sons, Inc.

U.S. Army. 2001. Manpower and Personnel Integration (MANPRINT) in the System Acquisition Process. Army Regulation 602-2. 1 June 2001. Washington, D.C.: U.S. Department of the Army.

U.S. Army. 2009. US Army FY09 Human Systems Integration Plan (Annex to the OSD HSI Management Plan), Version 1.0. Arlington, Va.: Headquarters, Department of the Army, G-1. Available online www.acq.osd.mil/se/docs/Army-FY09-HSI-Plan.pdf. Accessed March 14, 2013.

U.S. Army. 2012. The MANPRINT Program. Seven MANPRINT Domains. Available online http://www.manprint.army.mil/manprint/domains.html. Accessed March 22, 2013.

4

Achieving Overmatch

During its information gathering process, the committee identified many opportunities to improve the capability of dismounted tactical small units (TSUs) in ways that could potentially contribute to ensuring that future TSUs have decisive overmatch across all the tasks and missions described in Chapter 2. In the committee's judgment, many of these opportunities will have their greatest effect *only if* both materiel and non-material factors from across the Doctrine, Organization, Training, Materiel, Leadership and Education, Personnel and Facilities (DOTMLPF) domains are integrated in an optimized "capability solution," in accordance with the four overarching recommendations presented in Chapter 3. Such opportunities, or *capability options*, have interactive consequences, positive and negative, that will require the rigorous assessment and design approach described in Chapter 3 to find the best set. These interactive consequences often extend across two, three, or more of the operational capability categories discussed in Chapter 2: situational awareness, military effects, maneuverability, sustainability, and survivability. Thus, a set of capability options covering all five categories will typically need to be evaluated together during rigorous systems engineering (see Recommendation 2). Even so, the committee found that the most promising options, at least in terms of costs and payoffs the committee could evaluate, given its limited assessment base, can be roughly categorized into five high-priority capability-improvement areas:

- Dcsigning the TSU
- Focusing on TSU Training
- Integrating the TSU into Army Networks
- Balancing TSU Maneuverability, Military Effects, and Survivability
- Leveraging Advances in Portable Power

In each of these high-priority improvement areas, there are various options to consider and integrate, including improved or alternative technology options as well as non-materiel improvements in organization, doctrine, and other options that fall within the committee's definition of the human dimension. Work on many of the technology options is already in progress. Rather than make specific recommendations on which options are most worthy, this chapter explains why certain of the options are most likely to achieve overmatch following rigorous systems-oriented assessment.

DESIGNING THE TSU

The principles for achieving overmatch discussed in Chapter 3 would allow the Army to leverage Soldier performance as never before and to determine what TSU design would be most dominant across the full range of combat and stability operations.. A systems approach would focus on developing the metrics and opening the TSU design options to incorporate the full capabilities of Soldiers and equipment.

Ideally, the TSU would be viewed as a system-of-systems and not merely as an organization or formation. A proper system-of-systems analysis would then be able to determine design parameters for the optimal size (number of Soldiers), organization (number of fire teams, duties), and equipment (communication, lethality systems, etc.) of the TSU. The lack of published U.S. Army Training and Doctrine Command (TRADOC) documentation (e.g., an initial capabilities document or capabilities development document) that could help guide TSU design is a definite handicap. Although some future TSU missions may be similar, the tactics, techniques, and procedures (TTPs) that have been effective at squad level in Iraq and Afghanistan are unlikely to provide overmatch capabilities against all future adversaries.

Since World War I, the size of the Army squad has varied from 8 to 12 men. In the same time, the Marine Corps squad has been relatively stable at 13 men using three fire teams, except for a short period in the late 1970s when a Marine squad consisted of only 11 men. Army squad organization and size has been studied and reconsidered many times since World War II, starting with a 1946 infantry conference held at Ft. Benning, Georgia, and continuing at least through a 1998 study at the time that the light infantry brigade was being reorganized (Melody, 1990; Hughes, 1994; Rainey, 1998). The recommendations on squad organization and size in these studies flow from underlying functional factors assessed by the authors or study participants: recent deployment experiences; expected future squad missions; available equipment (e.g., weapons and communications), and non-materiel factors such as doctrine, leadership, and tactics. In short, squad organization and size have always been viewed as following from underlying factors, and the objective of the assessment has always been to improve future small unit performance in expected conditions of deployment. Given more-recent Army experience (deployments in Iraq and Afghanistan) with extensive use of dismounted small units conducting missions independently, the change in expected TSU missions under a wider range of military operations (i.e., stability tasks as well as offensive/defensive tasks; see Appendix D), and changes in available and emerging equipment to carry out these missions, the Army needs to conduct a new round of TSU organization analysis unconstrained by assumptions about numbers of Soldiers per unit, roles of unit members, etc.

The current TSU organization is not necessarily optimal. For example, it is conceivable that three fire teams (two rifle fire teams and one machine gun/grenadier/XM25 fire team) with appropriate changes in TTPs may provide significant improvements in maneuverability, military effects (particularly lethal effects), and survivability over the current two identical fire teams. The interplay among factors such as TSU size and organization with other DOTMLPF options is discussed further below, under the "*Organization*" heading in the section titled "Selected DOTMLPF Opportunities for Balancing Maneuverability, Military Effects, and Survivability."

TSU Design Considerations

The lack of an analytical foundation for squad performance limits future advances in capability to what infantry leadership is advocating at a given time; it precludes development of a stable TSU architecture. While the TRADOC Analysis Center has some force-on-force models that could be used, it has not used these to develop measures of performance (MOPs) or measures of effectiveness (MOEs) for the TSU. MOPs and MOEs for the current squad are based on operational experiments to assess particular materiel systems using scenarios developed for the experiment.

The Army has adopted 72 hours as the mission-duration standard for squad performance. A standard operation would require each Soldier to carry a "sustainment" load of about 60 pounds for the 72-hour mission, in addition to an assault load of about 45 pounds. This standard represents a "worst case" load, in the sense that a mission duration less than 72 hours would reduce the Soldier load. As a consequence of the 72-hour standard, Army developers have pursued multiple alternatives for manned and unmanned support vehicles, such as the M274 mechanical mule and the planned Soldier Mission Support System.

Robotic augmentation of TSU functions is a design consideration of enormous potential for Soldier and TSU capabilities in the future. However, proponents and developers of support vehicles for the squad continue to ignore the need to address many basic shortcomings that have been identified using prototypes, including several issues relating directly to TSU design. These include such things as: provisions for operator and maintainer manpower; vehicle mobility that is less than that of dismounted Soldiers (which means the vehicle cannot keep up with dismounts in complex terrain); load security when separated from the TSU formation, as well as other "minder" distractions; safety of dismounted Soldiers; and tactical noise and other signatures.

In addition to support vehicles to assist with load-carrying, there are portable unmanned aerial vehicles for reconnaissance, "port bots" for special purposes, and exoskeleton systems to consider in future designs for the most capable TSU organization. Appendix H provides descriptions of current relevant programs in robotics technologies.

As discussed in Chapter 3, until the Army develops a better understanding of TSU requirements, it will have no choice but to continue using worst-case approaches and faulty support concepts. The Army has a mature set of metrics for Armored Systems and Mounted Combat, which, together with models and simulations, can predict or estimate engagement, battle, and campaign outcomes for a given set of performance data and conditions. Analogous capability is needed for designing and evaluating dismounted TSU concepts. Using foundations developed in the 1980s, objective metrics, as recommended in Chapter 3, can be developed for social processes that are critical to achieving decisive overmatch, even if the scores on some metrics are not necessarily on an ordinal scale (that is, they are not ranked from a highest to lowest score). It should be possible for the Army to develop metrics for the dismounted Soldier and TSU in operations such as direct fire, movement, indirect fire coordination, information collection, mission planning, culturally aware behavior, situational awareness and understanding, and decision-making. These metrics, in the form of MOPs and MOEs, could be used in the near term as a basis for establishing realistic goals for future capabilities, as well as setting acquisition objectives and training readiness standards.

Development and analyses of TSU options will require collaboration among multiple Army activities, including the TRADOC Infantry School at the Maneuver Center of Excellence and the TRADOC Analysis Center, the Army Research Laboratory (ARL), Army Research Institute of Environmental Medicine (USARIEM), Army Research Institute for the Behavioral and Social Sciences (ARI), the Army Materiel Systems Analysis Activity, and the Army program executive offices for Simulation, Training, and Instrumentation (PEO STRI) and Soldier (PEO Soldier).

Finding: The task of developing metrics for the Soldier and TSU lacks organizational focus and responsibility. A single organization should have the responsibility for developing the metrics for dismounted Soldier and TSU operations.

Recommendation 5: The Army should transform and sustain the design of the TSU, including re-assessing unit organization and size, by the following actions:

a. Develop representative measures of performance (MOPs) and measures of effectiveness (MOEs) for the primary dimensions of TSU performance, and ensure these measures incorporate human dimension criteria.
b. Assemble a consortium of stakeholders to implement iterative work-centered analyses of the Soldier task workload and the TSU and Soldier-system performance required by increasing the scope (range, quality, thresholds) of TSU MOPs and MOEs. The analyses should enable development of predictive analytical models of Soldier physical and cognitive task and mobility performance, Soldier-to-Soldier task and mobility interaction within a TSU network, and TSU task and mobility performance.
c. Expand the TSU task and mobility model to predict influences of weapons, information collection, and information technologies on TSU MOPs and MOEs.

Such a TSU task and mobility model could be expanded in the mid-term to include individual Soldier and TSU social network factors as well as training states.

Soldier Performance

Changes in TSU design will require not only considerations for future missions and equipment but also adequate attention to the human Soldiers. Capabilities of the TSU and of the Soldiers in it are highly dependent on each other. Enhancements to TSU performance and effectiveness should also enhance performance and effectiveness of the individual Soldier. Likewise, Soldier enhancements should increase the performance and effectiveness of the TSU. For example, sharing situational awareness within the TSU enhances an individual Soldier's situational awareness. Enhancing the shooting skills of one Soldier will, in turn, enhance the lethality of the TSU. Future capability enhancements to the TSU and individual Soldiers should be designed to provide a synergistic effect that is greater than the sum of incremental improvement from each enhancement by itself.

As the Army considers encouraging enlisted careers reaching beyond the 20-25 years now the nominal standard, a shift in the expertise and experience levels of individual Soldiers might well have profound results on TSU performance, allowing the Army to capitalize on the training and experience of longer-serving deployment veterans.

In listening to and questioning Soldiers, troop leaders, and materiel designers, the committee learned that what is broadly known in the research and development (R&D) community about human physiological performance applicable to TSU dominance is not being applied by the Army. This deficit in applying critical information to understand and improve Soldier performance is discussed in the sections that follow.

Physiological Readiness

Most accept that sleep loss or extreme heat will affect physical performance. Less accepted, but well established in research, is that cognitive performance is just as profoundly affected by lack of sleep, temperature extremes, time zone shifts, poor nutrition, and extreme elevation changes. In particular, cognitive ability declines substantially with sleep loss (Miller et al., 2011; Thomas et al., 2000; Van Dongen et al., 2004). Depending on the individual, performance decline due to lack of sleep can be as much as 1 percent per hour after last rest. So, Soldiers operating 24 hours without sleep, assuming they were fully rested at the start of the 24 hours, may be operating with as much as a 24 percent cognitive deficit.

Seventeen to nineteen hours without sleep, which many consider not much more than a long day, can have the same impairment as alcohol consumption at the legal standard for driving under the influence (Williamson and Feyer, 2000). While there is wide variation among individuals in performance decrement from sleep deprivation, there is no correlation between individual self-assessments of their "alertness" and measured performance (Van Dongen et al., 2004).

As with raised blood-alcohol levels, "Can do!" spirit does not restore brain function, and it is the higher-order functions of judgment and analytical reasoning that fade first. Miller et al. (2011) discussed how "…sleep deprived leaders appear to have a diminished capacity to recognize their own sleep debt, as well as the sleep debt of their subordinates." The researchers surveyed recently returned combat veterans attending the Army Infantry Officers Advanced Course. Nearly 70 percent reported that their superiors received less or much less sleep than needed, 55 percent reported they themselves received less or much less than needed, and 47 percent reported their subordinates received less or much less sleep than required. The veterans noted that they averaged 4 hours of sleep per night during the periods of high operational tempo (OPTEMPO) that made up almost half of their time deployed.

The mental abilities required to achieve success exploiting network-centric capabilities are those most vulnerable to battlefield stressors that include sleep loss, environmental extremes, dehydration, and high OPTEMPO. Functional brain imaging studies show that sleep loss selectively deactivates the prefrontal cortex, the brain region where anticipation, planning, and situational awareness culminate (Thomas, et al., 2000; Wesensten et al., 2005a).

While the research literature on performance losses from a degraded physiological state is fairly robust, the committee found mention of these degradation factors missing in small unit leaders' considerations of operations planning. The mission planning aid described later in this chapter (see Recommendation 14 and preceding text) would be a tool for delivering this knowledge to small-unit leaders for operations and mission planning.

Most of the research on the physiological bases of degraded performance has concentrated on single-attribute relationships. Additional research is needed to understand the relationships among multiple degrading factors, such as the effects on physical, cognitive, and emotional performance attributes of combinations of sleep loss, poor nutrition, poor hydration, temperature extremes, exposure to extreme motion (air and ground vehicles), high elevations, and prolonged physical fatigue. Such research could better quantify the relations between the degrading factors and performance attributes relevant to mission planning, predictive simulations, and the models used for analyzing alternatives. Further, there is an equally urgent need for research evaluation of training, pharmacologic, and heating and cooling mitigation strategies, to include both the short-term and long-term effects of a mitigation strategy on Soldier fitness and Soldier health. For example, both Ritalin (methylphenidate) and modafinil are in some use by the U.S. military as antidotes to sleep loss, but little is known of the effects of such use on cognitive or emotional performance (Wesensten et al., 2005b). A second objective in this research should be to develop biomarkers that could indicate to TSU and other small unit leaders the physiological readiness of their Soldiers. Even when the OPTEMPO requires the assignment of Soldiers with reduced physiological performance, it is critical that mission planners understand the decrement in performance their TSU may encounter. A more complex third objective would be to learn how Soldiers differ in their sensitivity to the performance degradation factors and if such a sensitivity might be the basis for selection measures.

In the near term, physiological readiness could be inserted as a "must-do" checklist in TSU mission planning: "Have the Soldiers had a night's rest?" or "Is there an extreme elevation change planned?" And so on. When such precautions are not possible, both the assignment of squads to particular tasks and the number of squads allocated to a task should reflect a quantitative knowledge on the part of mission planners of the expected physiological efficiency of each unit.

Emotion Regulation

Small unit leaders reported that, on occasion, they had seen peer leaders perform while influenced by an emotional state brought about by family or domestic issues from home, by recent casualties, or other sources. Training should be incorporated into courses for small unit leaders to make them aware of the need for "mindfulness" in their decision-making and troop leading. This training could take the form of game scenarios that highlight the role of emotion regulation in tactical decision-making. Doctrine should be developed to guide leaders sensing the potential for emotion-driven degraded leadership in others or themselves. Research would be needed to develop effective measures of leaders' emotional states and the relationship to decision-making in high-stress and extreme environments, the effectiveness of the developed doctrine, and the effectiveness of leader emotion-regulation training. Research should also explore the potential for neurosensing of the emotional state of Soldiers and their leaders.

Resilient Soldiers

In parallel, perhaps, with the research on emotion regulation for leaders, research should seek to determine the attributes of resilience in Soldiers: the ability to perform effectively throughout the extremes expected in unified land operations. Such extremes are likely to include "three-block wars"—ranging from lethal fire and maneuver to humanitarian assistance and back to lethal fire and maneuver in a span of minutes. Increased resilience could also make Soldiers and units more survivable, both physically—able to survive threats posed by the enemy and the environment—and mentally—able to resist depression and assaults on cognitive ability, such as post traumatic stress disorder. Army research can provide new knowledge applicable to selection, assignment, and training strategies for increasing the levels of Soldier and TSU resilience.

The Army Center for Enhanced Performance, originally an enhanced performance program at West Point, has grown to have 100-plus affiliated professionals; it provides direction for basic training and interventive training events to several Army units. If institutionalized with Department of Army support, the center's results could be applied to expanded R&D program efforts in small unit leadership and small group social dynamics at ARI, the U.S. Army Medical Research and Materiel Command, the Human Research and Engineering Directorate (HRED) at ARL, and elsewhere.

Optimizing the TSU Social Network

Each TSU forms a discrete social network that must function efficiently. But little is known beyond the intuitive level about how the social network is forged within a unit, how it is maintained, and how personalities influence that process. For example, cohort training, in which Soldiers continue training and serving with the same unit beyond Basic Combat Training and Advanced Individual Training, showed some success in experiments during the mid- to late-1980s. The concept of keeping dismounted infantry units together for training and service over extended periods thus has merit, but its potential still needs to be objectively evaluated with other options (including combinations of such options), such as the master trainer concept discussed below.

As Soldiers move toward more interactions in electronic forums such as chat rooms, Facebook, and text messaging systems, it should be possible to automate the monitoring of each squad as an effective social network. This could yield huge benefits at low cost, if commanders were able to easily identify squads with degrading social cohesion or shifting internal loyalties. Such measurements lie well within the scope of current behavioral research and should be explored.

Soldier Selection

Selection figures prominently in achieving overmatch inasmuch as selection processes form the basis for recruitment, assignment, training, and retention decisions. Were one to succeed in defining the optimal squad or squads at a structural level, then it would also be possible to develop tools for maximizing the efficiency of individual squads during the process of personnel assignment. It is no secret that squads vary in their effectiveness and that this variation reflects the interaction of the personnel that make up each squad. While there has been a significant effort to enhance leader training and development at the squad level, there has been very little effort devoted to the notion that the squad—as a group of interacting personalities—forms a network that must function optimally if the squad is to achieve its potential. Just as commanders can improve the performance of individual squads by transferring squad members to different squads to overcome personality conflicts, it is possible that cohesive TSUs whose members work well together can be constructed based on achieving a proper mix of personalities.

Significant numbers of Soldiers are required for the volunteer Army, and the infantry specialties have traditionally not been the most selective. As a consequence, small unit leaders reported to the committee that, especially after first combat, on the order of 30 percent of their Soldiers were no longer effective and were thus a drain on the small unit for the remainder of the deployment. Although the Army's selection and placement process, Tier One Performance Screen (TOPS), is improving the prediction of initial training completion and first enlistment retention program, the selection technology could be further developed to learn if a propensity to become combat ineffective, at least in the perception of the small unit leader, can be predicted. Whether the outcome is an expansion of the Tailored Adaptive Personality Assessment System (TAPAS), another off-the-shelf or new psychological instrument, or a neuroscience-based biomarker, the need is critical if TSUs are to be more effective. Careful research on this theme might also reveal the role of leader traits and TSU composition on a Soldier being informally classified by a leader as combat ineffective.

Individual Differences

In all of the committee's data collection visits, a theme heard from the training base as well as from recently deployed Soldiers is that, through concentrating on the human dimension, the Army could exploit the talents and abilities of Soldiers to get closer to excellent performance, rather than settling for the lowest common denominator performance that was acceptable in the Cold War era. Using individual differences as a future force multiplier was an overarching recommendation in the recent National Research Council study, *Opportunities in Neuroscience for Future Army Applications* (NRC, 2009, Pp. 103-104):

> **Conclusion 17:** Neuroscience is establishing the role that neural structures play in the individual variability observed in cognition, memory, learning behaviors, resilience to stressors, and decision-making strategies and styles. Differences from one soldier to the next have consequences for most of the Army applications discussed in this report. Individual variability influences operational readiness and the ability of military units to perform assigned tasks optimally, but it is in many ways at odds with the conventional approach of training soldiers to be interchangeable components of a unit.
>
> **Recommendation 17:** Using insights from neuroscience on the sources and characteristics of individual variability, the Army should consider how to take advantage of variability rather than ignoring it or attempting to eliminate it from a soldier's behavior patterns in performing assigned tasks. The goal should be to seek ways to use individual variability to improve unit readiness and performance.

Exploiting the talents and abilities of individuals would apply to Army recruiting and training across the board, not just to dismounted infantry. But taking advantage of these individual differences at the level of dismounted TSU operations has several facets.

The Army Physical Fitness Test (APFT), although undergoing change, is an objective measure of physical readiness recognized as being combat relevant. Passing the APFT is a critical milestone for recruits to become qualified in a Military Occupational Specialty. Recycling trainees in initial entry training to give them more time to reach the fitness goals has been a constant feature of Army training. A recruit scheduled for Basic Training is also scheduled for follow-on Advanced Individual Training. With each Basic Training recycle, a follow-on Advanced Individual Training "seat" is vacated, resulting in wasted training resources. (Similarly, a trainee recycled in One-Station Unit Training "vacates" the remainder of his training seat.) A predictive screen for application in Military Entrance Processing Stations that would predict APFT potential success at the outset should be developed and used to schedule a recruit's Basic Training and Advanced Individual Training, or One-Station Unit Training, with or without physical training remediation. This would reduce Army training expenses for trainees, transients, holdees, and students and possibly improve trainee morale and retention. Fielding such a measure would be a needed administrative precedent for exploiting other facets of individual differences.

It is very likely that the differences in cognitive abilities and temperament among Soldiers exceed differences in physical appearance or ability. The Armed Forces Qualification Test, a subset of the Armed Services Vocational Aptitude Battery (ASVAB), is a respected measure of cognitive ability, and the ASVAB as a whole is a useful indicator of vocational affinity. These instruments are used to make both accession decisions and assignments for Military Occupational Specialty training. ARI developed TAPAS, which is another accession decision instrument that assesses temperament and interests. The use of TAPAS in a battery along with ASVAB and TOPS (the latter to assess educational attainment) is significantly improving training completion rates. This R&D should be broadened to determine if instruments deliverable in the Military Entrance Processing Stations could usefully predict learning styles, which could yield training base economies. To be successful, this effort would need to develop alternative training course syllabi for trainees with a preferred learning style, to capitalize on the potential for reduced training times and reduced recycle times or attrition.

Company and battalion commanders affirm that the difference in performance among squads (and among platoons) is large, suggesting factors of two or three or more between the lowest and highest performing units. Most of the research on small unit performance has focused on the leader as the central determinant of unit performance differences. The favorable fielding of TOPS provides a foundation for further and accelerated exploration of the role of team member attributes on collective team performance. This could then lead to research to define cognitive, noncognitive, and physical performance attributes that contribute to excellence in TSU performance.

If useful relationships are found among individual Soldier attributes and TSU performance, then further R&D could explore methods for filling TSU vacancies based on optimal complements to already assigned personnel already in a TSU. While using such data to make optimum assignments from a centralized authority may be beyond near-term feasibility within the Department of the Army, the potential for both avoiding poor TSU performance and for gaining broad excellence should not be entirely ignored. The personnel pipeline flow rates are sufficient that garrison or lower-level assignments could be made to attain most of the potential performance gains.

Finally, the Army may not be aware of much research, government-funded or otherwise, that could be highly relevant. For example, personality measures being studied for use by the National Aeronautics and Space Administration in assigning astronauts compatible for long-term missions to Mars may have utility in TSU assignments, perhaps for use as diagnostics in improving sub-par TSU performance or as markers for potential poor performance.

Changes in TSU design will require not just considerations for future missions and equipment but also adequate attention to the Soldier as a human. Capabilities of the TSU and of the Soldiers in it are highly dependent on each other. Enhancements to TSU performance and effectiveness should also enhance performance and effectiveness of the individual Soldier. Likewise, Soldier enhancements should increase the performance and effectiveness of the TSU. For example, situational awareness within the TSU enhances an individual Soldier's situational awareness. Enhancing the shooting skill of one Soldier will, in turn enhance the lethality of the TSU. Future enhancements to the TSU and Soldiers should be designed to provide a synergistic effect that is greater than the sum of incremental improvement from each enhancement by itself.

Several near-term actions support the goal of achieving decisive Soldier performance:

- Institutionalizing the functions of the Army Center for Enhanced Performance;
- Assembling a "Physiological Readiness Check List" for use in training and operational testing and refining development of nonintrusive physiological status monitors;
- Expanding research in the social processes of small units; and,
- Expanding research in individual differences, especially as applied to physical readiness screens used in recruitment and military training.

Recommendation 6: The Army should evaluate Soldier performance for the future mission effectiveness of the TSU in the near term by leveraging existing research and development and by considering all DOTMLPF (Doctrine, Organization, Training, Materiel, Leadership and Education, Personnel and Facilities) domains.

Mid to far term actions toward maintaining decisive levels of Soldier performance in TSUs include:

- Provide near-real time physiological readiness state reporting from Soldier and TSU to the command chain using physiological state monitors
- Leverage personality inventories, such as TAPAS, to determine the cognitive, noncognitive, and physical performance attributes that predict TSU performance.
- Conduct analyses to predict probable increased TSU MOP and MOE levels attainable if two-year and five-year technology goals are met and anticipated improvements are implemented.
- Explore the potential to discern the state of the social network, morale, and other performance-relevant attributes from the communications among the TSU members without invading individual privacy and without individual identifications.

Recommendation 7: To maintain the currency of representative measures for the primary dimensions of Soldier and TSU mission performance, the Army, including its doctrine and training, research and development, acquisition and testing elements, should undertake a recurring program (at least biannual) to re-evaluate Soldier performance considering the analytical foundation for the functional design of the TSU, including numbers of Soldiers, grades and specialties, career experience, organization, and external support requirements.

FOCUSING ON TSU TRAINING

Not only will Soldiers and TSUs be expected to do more, but an increased emphasis on exploiting human-dimension knowledge will demand innovative approaches to training. Focused training is essential to improving the performance of Soldiers and TSUs to levels that can ensure overmatch. Small unit training and leader training are more important than ever, not just because of sophisticated technology but because the TSU is the centerpiece of future Army operations.

The TSU Training Imperative

The TSU must have mastery of the methods, tactics, and technical knowledge and skills required to accomplish the assigned missions before the missions are undertaken. Current senior Army leader training emphases are that future missions comprising the entire range of military operations, including counterinsurgency and wide area security as well as combat operations, are to be expected; greater use must be made of the full spectrum of training technology, virtual, constructive, and real; and commanders have full responsibility for the state of training of their units.

TSU "Master" Trainer

To fully exploit available information technology, as well as maneuverability and military effects technologies, the TSU must have additional training resource coordination and leadership capacity to facilitate mastery-level TSU performance. The master TSU trainer, as envisioned by this committee, would be assigned at company or battalion level for continuity of training and rapid assimilation of new technology. Such trainers would serve in a dedicated training capacity and would have special qualifications in skill acquisition and learning. They would be key advisors to the commander on TSU matters. These trainers would not be in the normal operations planning role of the operations staff; they would have a role analogous to a sideline and practice coach, being sure the team fielded is fully prepared for the contests ahead. The master TSU trainers would not be expected to be TSU players. The Army's current "player-coach" model of unit training leadership is short of what is needed to exploit the mission command networks and systems and military effects equipment (including weapons) provided for TSU use. The master trainer would understand and employ the full potential of the training technologies available, wherever the company/battalion is posted or deployed.

The TSU master trainer would assess the strengths and weaknesses of each TSU in the company/battalion; understand the existing systems and new technologies available in the next readiness cycle; and prescribe a training syllabus to get each TSU to a mastery level on both current and forthcoming systems.

Attributes for TSU master trainers might include legitimate academic degrees in education or psychology, as well as TSU leadership experience at levels above the company. Subject to periodic performance review, the TSU master trainers might be in "tenured" positions similar to master recruiters in the U.S. Army Recruiting Command.

TSU Training Attributes

TSU training must be experiential (scenario-based) and situated in realistic environments (live, virtual, or constructive); it must mirror the complexity of real-world operational environments; it must be accessible when units are deployed as well as at their home station, and it must rapidly incorporate and share the recent experiences of other units in similar situations. While commanders may be responsible for their unit's training, subordinate leaders must be given the means to accomplish the training. That is, TSU leaders must have training support to enable the near-full-time training of their TSU with minimal TSU Soldier downtime while leaders "prepare the training." To be effective with respect to both cost and training transfer, the training architecture must take a holistic view of the TSU and TSU leadership; the training architecture, training technologies, and facilities; and the training support staff. This holistic view should enable faster, cheaper (more effective training per unit cost), and more individualized instruction.

TSU Training Objectives

The first step, the one on which all else depends, is to define the objectives of TSU training properly: define the skill requirements at a level of detail sufficient for developing training content and evaluating delivery means. TSU training content should include adaptability, which can be based in part on ARI work on training for adaptability. Cultural awareness and cross-cultural skills should be included, which can be based in part on work by ARI and the Marine Corps' "Operational Culture" approach. Training in social interaction—being a "good stranger" in a host culture—might build on the Strategic Social Interaction Modules program of the Defense Advanced Projects Agency (DARPA). The training objectives should reflect the TSU performance metrics (MOPs and MOEs) established in accordance with Recommendation 3 of this report and discussed in both Chapter 3 and preceding sections of this chapter.

Realistic Sociocultural Training

Achieving TSU training objectives for noncombat tasks in stability operations may require an increased level of fidelity in virtual and constructive training facilities to recreate complex sociocultural situations. Leaders and instructors will need training development and management tools for rapid construction of training scenarios, rapid and inexpensive translation of a deployed squad's real-world experiences into training for other squads, and tools to pick the best (most cost-effective) type of training environment and level of fidelity for the training objectives. Training environments may be anywhere along the "live-virtual-constructive" (L-V-C) continuum and may often combine elements of two or more of these training environments. In virtual and constructive training environments, more realistic synthetic entities, especially with methods to improve behavioral realism, are needed. This can be accomplished in numerous ways with current technology, as illustrated by the following examples.

Autonomous Conversational Characters. To enable the TSU to train on skills requiring interaction with the local population, the current generation of games and simulations would need to be upgraded to include autonomous characters capable of conversation. The TSU should be able to have bidirectional interaction with these characters, including both verbal and nonverbal forms of communication. The characters should provide a plausible cultural representation, which is an area requiring ongoing research in terms of verification and validation criteria for cultural fidelity. Such conversational characters would support a host of different training applications, ranging from providing counseling skills for leader development and cultural awareness training to providing conversational (language) support to dismounted Soldier training for wide area security competency. The conversational and social capabilities should eventually be integrated with the autonomous non-player characters, but this needs to be done in stages due to the complexities of the architectures required.

Social Simulators. A social simulator would model groups of people and their relationships to one another. It should show the potential second and third order effects of operations in the human terrain so that TSUs can see how a local contact or action may affect the population of an area over time. While not claiming to be predictive in nature, these models should provide plausible reactions to information, interaction, and operations in an area. A local interaction with one individual can potentially have an outsized effect on the network of relationships of that individual. As with the human representations of individuals, further research is required for how one would provide verification and validation of a social simulator illustrating the current state of the art. Successful validation of such social simulation models is described in a recent National Research Council Report (NRC, 2008) and by Louie and Carley (2008).

Massively Multiplayer Online Games (MMOG). Use of MMOG environments augmented with autonomous non-player characters can support operational training for both combat (offensive/defensive) tasks and stability tasks. Ideally, the MMOG would support the use of a social simulator. Assuming that the software meets requirements for use on Army installations, an MMOG can be used to support home station training, as well as training at institutions (e.g, Army schoolhouses) and in the deployed force. One such system, called EDGE, is now being developed by the U.S. Army Research, Development and Engineering Command.

Assessment Tools for TSU Training

Objective measures of performance fed back to the trainees make the difference between effective training and "busy-work." Technology for rapid unobtrusive performance data collection during L-V-C sessions must be built into the training technologies. Instrumenting virtual, constructive, and simulator-based training with machine learning algorithms can enable individualized, automated assessment of trainee performance and can be a useful aid for leaders and instructors. Near-real-time training feedback allows leaders to adapt training to the appropriate level for the TSU and for individuals, to accurately diagnosis performance deficits, and to increase the training challenge to steepen the slope of the learning curve. (Soldiers learn more quickly when feedback shows skills are being quickly acquired.) Machine learning for diagnosis and feedback would be a significant instructor tool. Use of neurophysiological measures to estimate "operator state" might improve the resolution of training assessment and individualization. Development and fielding of a robust training management system for small unit leaders with improved record keeping for analysis of training effectiveness and digital record keeping for individuals is an enduring priority. Current Army proposals and programs for training avatars have two components: (1) a facility for collecting and exploiting for training management the training performance data on individual Soldiers, and (2) graphical representation of the individual's performance attributes in virtual environments. Research may indicate virtual avatars linked to actual performance would be a useful feature in virtual training environments, but the benefits of the graphical representation will be unknown until empirically demonstrated, which itself will be complex.

Simulation Technology and Devices for Automated Training

The benefit of automated tools for training management of small units is unquestioned, having been well demonstrated by ARI since the 1980s, and it could proceed without waiting for the virtual representation to be proven. Appendix F discusses two areas where improvements relevant to Army training needs could accelerate and accentuate the effectiveness of training through the following simulation technologies, which span the L-V-C training spectrum:

- **Authoring Tools.** The cost of scenario authoring is a leading limit to more pervasive and practical use of integrated L-V-C training. Advances in scenario authoring tools are needed that reduce the cost of developing new training scenarios to meet new operational requirements and operational environments.
- **Tools for Immersion Training.** Trainee immersion, so critical for the L-V-C integration concept, is especially challenging for application to dismounted TSU operations, where each Soldier should experience a unique, perceptually realistic relationship with the immediate terrain. Realistic simulation of walking, running, crawling, and taking cover—all routine TSU member behaviors—are technology challenges to be overcome to make the immersion experience valuable for Soldier training.

Instructor Training

In one of two training sites visited by the committee, technology was installed to provide trainees fall-of-shot feedback in basic rifle marksmanship; however, it was not being used for more than a modest fraction of its potential. The feedback system appeared to be used more as a means of keeping non-firing trainees occupied than as a training assist. Improved instructor training would ensure understanding of the important role of feedback in learning.

Adaptive and Accelerated Training

Training that is tailored to an individual's progress is in widespread use in maintenance and other technical training in the Army and in the other services. A fresh look at the constituent skills needed by a TSU Soldier might reveal areas in which this approach could be usefully applied, although this fresh look would need to include variances from the current lock-step One-Station Unit Training structure. Technologies to aid adaptive training with physiological measures are coming from psychology and neuroscience and are approaching commercial availability for consumers. Machine learning could simplify the results for instructors and speed and focus the results for trainees. Using nonveridical feedback in virtual environments could accelerate the speed and accuracy of training for initial entry and skill sustainment training of units. In virtual or other scenario training, providing trainees explicit comparison between seemingly similar (or dissimilar) situations would enable development of abstracted representations of situations. This builds robust, flexible knowledge bases that afford transfer to new situations. Other adaptive training techniques include the following:

After Action Review (AAR) Systems for Squad Operations. AAR is an essential step in the training process (Meliza et al., 2007), and AAR systems have been shown to increase the effectiveness of learning considerably (Katz et al., 2000; Katz et al., 2003; Schurig et al., 2011). To meet the TSU training imperative, all simulations and games used for training should have an automated AAR system included as a standard part of the system. The AAR shows detailed cause and effect in both offensive/defensive and stability operations so that a squad can review what it did right and what it did wrong; it also suggests ways of improving. The AAR should be linked directly back to the learning objectives and assessment tools populated by the authoring tools described above.

Adaptive Tutoring. Many studies have shown the effectiveness of personalized tutoring (Fletcher, 2011; Bloom, 1984). The differences in learning between standard lecture-based learning and personalized tutoring are dramatic. To accelerate and accentuate the effectiveness of TSU training, a significant focus should be placed on developing adaptive training systems that model and assess the user and consequently personalize the learning experience by providing tailored feedback and instruction. To accomplish this, the system will continually assess the state of the learner, including physiological monitoring as well as knowledge and skill assessment. It will provide feedback and tutoring as well as motivation, and it will adapt the pace and content of the instruction to optimize the learning path for the individual. As has already been pointed out, the feedback for such systems can be tied back to the learning objectives and the authoring of content for the system.

Mobile Learning Applications. One of the barriers to training is access. It is limiting to think of training occurring only when the Soldier or TSU is in a classroom, a simulation facility, or a training area. It is now possible to bring training to the TSU wherever it is, through the use of mobile devices. It will be possible to deliver standard instruction not only through web portals but also on smart phones and digital notepads. As autonomous conversational characters are ported onto mobile devices, it will become possible to train on human dimension skills such as negotiation, counseling, and building trust. Human terrain applications will enable greater effectiveness in the sociocultural dimension of the mission set.

Nutrition Training

Currently, initial entry training includes just 45 minutes on general nutrition and health. For TSU Soldiers, this initial training must include the effects on cognitive as well as physical performance of nutrition, hydration, sleep, dietary supplements, tobacco, and alcohol, as well as food and water hygiene safety. Training should cover both the acute—next-hour and next-day—effects and the chronic effects that occur over months and years. Nutrition, hydration, and other life style choice lapses could be built into the Army's first-person game[1] distributed to recruits and Soldiers.

As part "team manager" and part "team captain," the TSU leader must lead by example and by counseling the other TSU members on the physical and cognitive performance effects of nutrition, hydration, sleep, dietary supplements, tobacco, and alcohol. Researchers and scientists maintained that lack of positive leadership led to Soldiers mis-use of available rations. Medical researchers reported that high dysentery rates for units deployed in Iraq and Afghanistan were largely attributable to lack of ration discipline.

Findings and Recommendations on TSU Training

Finding: To achieve overmatch, Soldier and small unit training will have to emphasize both physical and cognitive performance, especially in areas of leadership, physical and cognitive fitness and resiliency, aptitude in human and social-cultural awareness, and ability to perform under severe stress from combat, information overload, physical demands, weather, severe temperatures, etc.

The development of training objectives has a rich history in the training literature. The science of work analysis has struggled for many years to define training objectives in a way that is measurable; has face validity; and considers the individual's knowledge, skills, and abilities and the demands of the specific job to be trained (Wilson et al., 2012). Cognitive task analysis methods have proven useful in interviewing experts, extracting key knowledge, and identifying learning objectives (Clark et al., 2008; Crandall et al., 2006). More recently, Mission Essential Competencies have been defined as a unique approach for training analysis in military settings. The process for developing Mission Essential Competencies is both task- and worker-oriented, representing a blended job analysis approach for understanding the requirements of the job (Bennett et al., in press; Garrity et al., 2012; Alliger et al., 2012).

Recommendation 8: The Army should focus training for the individual Soldier and TSU in the near term as follows:

- Define TSU training objectives to produce TSUs that perform acceptably on the TSU MOPs and MOEs.
- Produce nonintrusive physiological status monitors to allow self-awareness and command chain assessments.
- Apply results of research in individual differences to the administration of TSU training.

---

[1]In video games, "first-person" refers to a graphical perspective rendered from the viewpoint of the player character. Perhaps the most notable genre to make use of this device is the first-person shooter, where the graphical perspective has an immense impact on game play.

- Expand sociocultural training capabilities to produce necessary TSU skills within time and resource constraints expected for TSU deployments.
- Expand instructor development to incorporate current theories of learning and feedback.
- Develop a concept for TSU master trainers to be assigned to company or battalion level to ensure continuous effective training of TSUs.
- Develop tools for TSU leaders (and leaders at higher levels) to assess Soldier and TSU training readiness against the TSU MOPs and MOEs.
- Ensure that effects of nutrition, hydration, sleep, dietary supplements, tobacco, and alcohol on cognitive and physical performance are incorporated in all modes of training of Soldiers and noncommissioned officers, including electronic games as well as live, virtual, and constructive simulations for individual (self) and group training.

Recommendation 9: In the mid to far terms, the Army should refine its focus on training for the individual Soldier and TSU by increasing the resolution of its suite of assessment tools to allow tracking of Soldier and TSU skill acquisition through and during each individual and collective training event, including live, virtual, and constructive simulations and electronic games.

INTEGRATING THE TSU INTO ARMY NETWORKS

Soldiers and TSUs currently have limited organic capability (e.g., radios) to integrate maneuver and fires in all environments to achieve tactical overmatch. They must have "reach back" and "reach forward" capability in the areas of mission command, intelligence, fires, mission planning, location/tracking of forces, social networks, and all of their associated enablers.

Based on the approaches used in recent years, the Army believes integration can be achieved by providing Soldiers and TSUs with a geolocation system and radio-enabled information systems that are integrated into current and evolving Army networks, such as the Warfighter Information Network-Tactical. This perception arises in large part from Army development programs: Future Force Warrior (part of the canceled Future Combat Systems), Land Warrior (canceled program), and the more current Nett Warrior (formerly known as Ground Soldier System). All of these programs focused on providing Soldiers a Global Positioning System (GPS)-based personal location system combined with data communications that enable a Soldier to view his location, the location of other Blue force personnel and vehicles, information on enemy spot reports, mission command graphics, text messages, and similar warfighting information. The goal of these Army systems was to enable situational awareness at the Soldier level, not necessarily to integrate the Soldier within the small unit.

Definition of Network Integration

Network integration includes the development of dynamic communications, information, and socio-cognitive networks as well as associated enhancements in the DOTMLPF domains. The capability needs associated with these three types of networks were described in detail in Chapter 2 and are summarized here:

Communications Networks. For communications networks, advances are needed in hardware, frequency spectrum (particularly for bandwidth rates), and user interfaces. The Army is attempting to address these needs with the Nett Warrior Program, which is experimenting with smart phones, leveraging the technology comfort of Soldiers. However, the low-bandwidth spectrum currently available at the TSU level limits its use, making the system dependent on commercial cellular networks. At the TSU level, high bandwidth rate communications networks are needed that can operate in austere locations, in complex terrain (e.g., urban or mountainous), in all weather, and can overcome cyber security threats in the tactical environment. Night operations require communications devices whose light and noise do not compromise one's security and are usable with night vision devices.

Information Networks. Information networks provide TSUs with access to a variety of databases and sensors (ground and air, manned and unmanned). Capabilities should provide access to both internal/organic (assigned to the TSU) and external information sources, such as the Tactical Ground Reporting Network (TiGRNET), brigade databases, or streaming video from an unmanned aerial vehicle assigned to battalion headquarters. Linkages to data information sources such as TiGRNET and battalion/company/platoon databases must allow the TSU and Soldier not only to access mission-relevant information but also to provide critical intelligence information as input—making every Soldier a sensor. Sensor networks should provide critical information on the identification, location, and tracking of friendly, enemy, and noncombatant personnel, especially in cluttered, urban environments where GPS signals are weakened or completely blocked. Sensors are needed that can sense through walls or on the other side of obstacles. Sensor missions organic to the TSU (requiring internal capability) are discussed briefly below. The design and development considerations for organic sensor technologies are discussed in the section on Network Integration Priorities, and a more extensive assessment of sensor capabilities and technology opportunities is provided in Appendix G.

Socio-Cognitive Networks. Dynamic socio-cognitive networks link to databases that can help characterize a person's community and identify his/her association with overlapping communities. A growing array of such networks can be used to identify and interact with local leaders and visualize social connections. Means to extend network access to Soldiers and TSU will increase knowledge and understanding of the human terrain.

Integration of the TSU into these three types of networks can only be achieved through concerted DOTMLPF efforts. In both combat and stability operations, all three networks support the Soldier's and TSU's ability to rapidly shape the operational environment before engagements by exploiting every aspect of the populace to their advantage, thereby helping to erode the threat from the noncombatant populace and to achieve minimal collateral damage or loss of noncombatants.

TSU Organic Sensor Capabilities

Both organic (internal to the TSU) and supporting (external) sensor capabilities are essential to TSU overmatch. The external information derived from sensor systems maintained by, and sensor data processing at, higher echelons comes to the TSU over its communications links with the larger Army network.[2]

The three general sensor mission categories at the TSU level are situational awareness, force protection, and precision targeting. Sensors providing situational awareness yield timely information about current events within the spatial proximity of the squad, such as locations of dismounted threats, approaching vehicles, or potential targets within buildings. Navigation sensors for use in a GPS-denied environment fall into this situational awareness category. Force protection sensors are used to provide adequate warning to minimize lethal engagements involving rockets, artillery, mortars, small arms fire, mines, improvised explosive devices, and chemical-biological-radioactive-nuclear agents. Precision targeting sensors provide fire-control information to Blue Force weapons; examples include infrared seekers or the counter-battery solution generated from weapons location radar. Electronic warfare sensors are an important fourth mission category for the Army generally, but for the dismounted TSU the principal organic sensor application in this area is for anti-jamming, which can be considered a form of electronic force protection for the TSU. Table G-2 in Appendix G, with the accompanying text, characterizes sensor tasks and technologies relevant to squad-level operations in these three sensor mission categories.

Potential Benefits

The integration of Soldiers and TSUs into the Army network would satisfy capability needs in all of the areas required to increase decisive overmatch, especially situational understanding, maneuverability, and survivability.

Situational Understanding

A crucial concept to guide this integration into the Army network is the necessity of ensuring that TSU leaders and individual Soldiers have sufficient *situational*

[2]The overview in this section, which is drawn from Appendix F, focuses on sensor needs organic to the dismounted TSU and how that sensor capability interacts with other TSU capabilities across the DOTMLPF domains. Appendix F reviews the capability needs and technology solutions for both organic and supporting sensing missions.

understanding. As discussed in Chapter 2, full situational understanding requires all three levels of situational awareness, namely:

- Level 1 situational awareness is the perception of disaggregate elements of information acquired from data received from sensors either directly or indirectly; plus
- Level 2 situational awareness, often referred to as situational understanding, is achieved when Level 1 perceptions are further combined, interpreted, stored or retained for use by a Soldier or TSU, plus
- Level 3 situational awareness is reached when Level 2 perceptions are applied to project possible future events and anticipate outcomes.

Coupling the network with geolocation sensors will provide a much-needed Blue Force tracking capability to locate and track not only members of the TSU but also adjacent units. The location of threat forces will be derived from the combination of access to sensors (including imaging and streaming video from robotic air and ground vehicles) as well as from spot reports and other intelligence data. Access to databases and systems (e.g., DARPA's Tactical Ground Reporting system, TIGR) will provide cultural data and other counterinsurgency-focused information. In addition to situational understanding being enhanced for each Soldier, the TSU will also benefit from enhanced shared situational understanding.

Military Effects

Enhanced situational awareness, the ability to rapidly transmit and receive tactical information (e.g., mission command graphics and fragmentation orders), access to intelligence organizations and lethal systems supporting the TSA, the ability to rapidly generate and access reports, enhanced capabilities to plan and rehearse missions, and improved ability to support on-the-spot training and rapid utilization of lessons learned will all contribute to Soldiers and TSUs making sound decisions on application of military effects and dominating both lethal and nonlethal engagements. As noted below in the section on "Balancing TSU Maneuverability, Military Effects, and Survivability," the lethal capability organic to a dismounted TSU must be enhanced with the ability to access, coordinate, and integrate joint fires (e.g., from company mortars through close air support) to suppress and destroy enemy targets, especially at extended ranges (those beyond capabilities of organic weapons), if the TSU is to have decisive overmatch in all combat missions. However, these essential joint fires are only fully capable of supporting the TSU if the TSU is always and continuously integrated into the network through which joint fires are requested.

Maneuverability

Rapid real-time access to maneuver-related mission command graphics, terrain information, and other maneuver-related information (e.g., location of obstacles) will greatly enhance Soldier and TSU mobility. Access via the network to supporting joint fires for both suppressing and engaging threat systems will also enhance overall maneuverability. Integration into logistics networks will enhance the ability to rapidly resupply a TSU when needed, thus reducing need to carry excessive ammunition, food, water, etc. and thereby contributing to reducing Soldier load.

Survivability and Sustainability

With enhanced situational understanding and maneuverability, the vulnerability of Soldiers and TSUs to threat systems and fratricide incidents should be significantly reduced. Improved access to medical evacuation information networks will assist in maintaining the lives of wounded warriors. Network-supported access to physiological data on individual Soldiers and TSUs will assist leaders in making better decisions with regard to resupply (water, nutrition, etc.) and rest cycles. Advanced warnings of weather extremes will also be helpful.

DOTMLPF Considerations

The Army has focused on materiel solutions to achieve the benefits of network integration, but all elements of DOTMLPF must be considered, especially doctrine, organization, training, and personnel.

Doctrine

Multiple doctrinal issues must be addressed to integrate the Soldier and squad into the network. Determining the critical information requirements for Soldiers and TSUs is the most critical issue. Some related work has been done by the ARL/HRED Field Element at Fort Benning, Georgia. However, much more needs to be done. The information requirements should also be identified by phases of a mission. For example, for an offensive mission, information requirements should be identified for the planning, pre-assault, assault, and consolidation as shown notionally in Figure 4-1. In the type of operation illustrated in Figure 4-1, enemy information is critical during all phases of the mission, whereas historical information is very important during planning but has little significance during the other phases of the mission. Similarly, different members of the TSU may have varying levels of need for information—e.g., the Squad Leader has a much greater need for information than a rifleman.[3]

Another doctrinal issue is the development of TTPs for utilizing these network capabilities. For example, in the Smart Sensor Web experiment conducted at Fort Benning in 2002, TTPs such as *remote reconnaissance*, *information overwatch*, and *watch my back* were developed by the participating Army and Marine infantry platoons. *Remote reconnaissance* involved the accessing of remote sensors in the objective area by

[3]The Army Science Board is currently reviewing information needs for squads in its study on "Data-to-Decisions." The committee believes this is an important complementary effort to this study.

Fire Team Leaders while the Platoon Leader and Squad Leaders were planning a mission. As Fire Team Leaders identified information pertinent to the impending mission, they fed it directly to their leaders as "real time" intelligence for planning the mission. *Information overwatch* exploited the fact that during a multipart mission (one with more than one squad objective) one squad is in the assault, one squad provides covering fire, and a third squad stands by to take over the assault or covering fire task in the next part of the mission. The stand-by squad is assigned the task of information overwatch—it digests the available tactical information and passes only critical information to the assault squad or covering fire squad, as appropriate. Finally, *watch my back* was a tactic involving a last-minute check with the sensor network just before entering to clear a building or room.

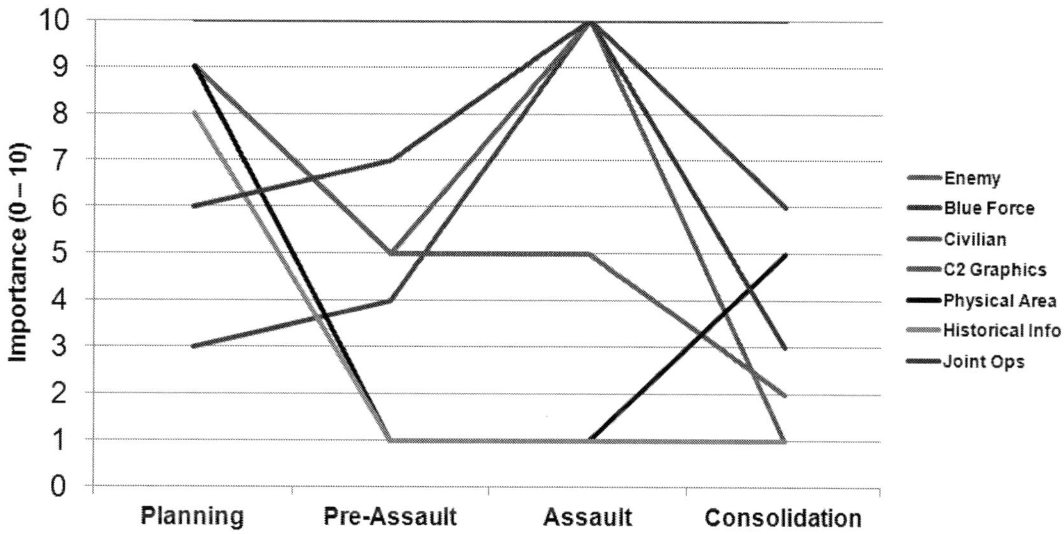

FIGURE 4-1 Notional information requirements for TSU with offensive mission.

New TTPs like these three are needed to take advantage of the capability of today's information systems at the TSU level, A good starting point on this has been made at the company level by providing an information overwatch team called the Company Intelligence Support Team (CIST).[4] As networks become more intelligent, and the function of fusing/assessing information and notifying Soldiers and TSUs becomes more and more automated, TTPs will need to evolve.

Additional TTPs, closely related to the organization of the TSU, will need to be established to address the degree of network connectivity for each member of the TSU. For example, if all members of the TSU receive radios, it must be determined which and when members of a TSU transmit and receive. For example, during an assault, riflemen may only receive while Fire Team Leaders and above will both transmit and receive. However, during the establishment of a defensive position, all members of the TSU may need to transmit and receive.

[4]Center for Army Lessons Learned (CALL) Publication 10-20, *Company Intelligence Support Team Handbook,* will assist leaders in understanding the mission and purpose of CISTs and how to better use these teams. This small-unit intelligence capability enables the company to maintain situational awareness and possibly even attain brief periods of situational understanding and information superiority.

There may also be a need to determine who talks to whom in a point-to-point tactical network. An example of such a TTP would give a Fire Team Leader the ability to communicate selectively with all members of his fire team, the other Fire Team Leader, the Squad Leader, or a combination of these options. The Squad Leader would be able to communicate selectively with all squad members or with just the Platoon Leader, the Platoon Sergeant, and/or the medic. This doctrinal issue may also be modified by unit Standard Operating Procedures.

TTPs will also need to address those capabilities a TSU should exploit—for example, the integration of both organic and external fires into the TSU's maneuver operations to defeat line-of-sight and non-line-of-sight threats that inhibit TSU movements. .For instance, will Squad Leaders be able to utilize communications and information networks to achieve effects from joint fires at ranges outside small arms range in austere environments?

Training

Obviously, new TTPs will require new training programs. Training equipment and facilities will need to be developed that stress the use of the new materiel, the new TTPs, the new organization, etc.

Computer simulations will need to be developed that include the use of information systems—for the computer-generated virtual Soldiers and TSUs as well as for the live trainee-participant's user interface. (For example, the trainee-participant may have a "radio" with which he can talk to the avatars of other networked live players, as well as with the computer-generated virtual players). Similarly, the computer-generated players will need to exhibit human behaviors that would be influenced by the use of dynamic communications, information, and socio-cognitive networks (for instance, the computer-generated players would need to exhibit varying levels of situational awareness).

Live training facilities would likewise need to be upgraded beyond being benign brick and mortar facilities; they would have both a realistic electromagnetic environment that could cause interference, as well as realistic building materials that might also interfere with the operational use of information systems at the TSU level. As an example, the current use of steel CONEX containers as buildings causes unrealistic radio propagation problems that can be detrimental to training. Training systems will also need to incorporate networked robotic systems.

Small unit leaders will need to be educated and trained on how to best exploit this evolving capability of being integrated into the network. Their professional development will need to address "reach back" and "reach forward" capability in the areas of mission command, intelligence, fires, mission planning, location/tracking of forces, social networks, and all of the associated enablers for these functions. Leaders will need to be not only tactically competent but also technically competent to adequately exploit the network.

There are undoubtedly numerous other doctrinal and TTP issues that will need to be addressed in order to integrate the Soldier and TSU into the Army network. TRADOC needs to begin exploring (e.g., through experimentation and rehearsal-of-concept drills) this new area of capabilities for the Soldier and TSU.

Organization

The Army has reacted to the need for exploiting the network at the company level with the organization of CISTs, but similar organizational changes must be considered at platoon and lower levels. For example, who within a TSU will be equipped with what information system? Most likely, every Soldier will be equipped with a geolocation sensor to provide his location. via the network, to leaders at the TSU level and higher. However, whether or not every member of the TSU needs a radio is questionable. There may well be situations where all members of the TSU need to transmit and receive, while there are other times when many members of the TSU need only receive. There will be need for appropriate support personnel at the platoon and company level, to include a team to assist TSUs with dealing with electronic warfare and cyber attacks. The integration of networked robotic systems into the TSU-level organization also needs to be addressed.

Personnel

From the perspective of today's Soldier and current personnel selection methods, one might ask two questions:

1. Are the current selection criteria for infantry Soldiers adequate to support the integration of Soldiers and TSUs into the network?
2. Are current Soldiers and TSUs ready to be integrated? With respect to this question, think about the difference in radio chatter one might expect in a firefight when comparing a highly trained Ranger TSU versus a regular Army TSU. Most likely the latter would have less transmission discipline. Is this because of training and experience alone, or does selection also play a role?

It is extremely important that Soldier-network interfaces (voice, digital, haptic, etc.) and the information being conveyed be designed to accommodate the skill levels of Soldiers and TSUs. A significant part of the overall systems engineering effort is to optimize the impact of these interfaces and accompanying information on Soldier and TSU performance and effectiveness.

There will also be a need for more information technology-savvy repairmen and software programmers at the company level. The latter will be needed to aid in rapid changes to systems to adjust to electronic warfare and cyber attacks. Related materiel considerations include improved frequency spectrum allocations for TSU networks, maintenance support and repair parts, and possible changed power requirements generated by these networked systems.

Network Integration Priorities

Materiel advances are needed in the technologies supporting the communications, information, and socio-cognitive networks. Advances in network science, sensors, system interfaces, and power systems are critical for future enhancements. The Army is addressing many supporting materiel developments.

Communications Networks

The current Nett Warrior program claims to be an integrated situational awareness system for the dismounted leader. The focus of the system is to graphically display the location of TSU leader and Soldier locations on a geo-referenced map image. A secure radio connects the display (currently similar to a smart phone) to other Net Warrior systems and the larger Army network. Access to the larger Army network provides higher echelon data and information products to assist in decision-making and development of situational understanding. Soldier position information will be available through the use of the Army Rifleman Radio (JTRS HMS Rifleman Radio (AN/PRC-154)).

As General Stan McChrystal once stated: "You don't give a senior leader a Blackberry or an iPhone and make them a digital leader."[5] The integration of information and geolocation systems does not in itself provide dominance and ensure optimal decision-making and situational awareness capabilities for the Soldier and TSU. For example, a network connection may assist with providing data for developing an individual Soldier's Level 1 situational awareness, but even the Level 1 awareness and especially Level 2 and 3 situational awareness will be most efficiently developed with strong human dimension enablers such as enhanced training and education, leadership development, shared knowledge and experiences, and qualified personnel.

What is truly needed is an integration of DOTMLPF enhancements in the areas of dynamic communications, information, and socio-cognitive networks with improvements in personal and local sensors. Examples include integration of sight and sound situational awareness inputs to the individual Soldier, information collection sensors on robotic platforms, biometric sensors for identifying civilians, and, sensors or other devices supporting the location and tracking of dismounted personnel and warfighting platforms.

Emphasis in the near term should be on developing organic communications capabilities with some access to adjacent units and immediate higher echelon organizations. Among potential materiel solutions are the following:

- Enhanced real-time, point-to-point, long range, high-bandwidth, non-line-of-sight communications. This capability is particularly important in complex and urban terrain, where transmission propagation is often severely degraded. However, improvements in range and other features typically increase power needs.

---

[5]Ackerman, Spencer; "Stan McChrystal's Very Human Wired War;" Wired; January 26, 2011; available online at www.wired.com/dangerroom/2011/01/stan-mcchrystals-very-human-wired-war/.

Consider the use of relay systems—either within the radios (similar to current vehicular systems) or through the use of relay points (especially utilizing robotic systems). Bandwidth rate issues can be addressed either by assigning higher frequencies to the TSU (a low-tech approach) or by developing technologies to manipulate the frequencies available to provide bandwidth rate improvements.

- Radios that can be remotely manipulated by leaders. In this case, leaders can determine and control remotely who transmits/receives and who only receives during various phases of a mission.
- TSU-level network management systems that provide the ability to switch from broadcast transmissions to point-to-point protocols to set up tactical social networks. With such systems, TSU leaders can, with the flick of a button, determine whether they are talking to an individual or to selected groups of individuals—below, at peer level, and above.
- Hands-free interfaces that require minimum time for accessing and sharing information. It is important for Soldiers and TSUs to focus their attention on assigned tasks, the mission, and the objective. Interfaces (especially wearable, lightweight screen displays; and voice, gesture and haptic interfaces) must be designed to quickly and efficiently convey and collect information to/from individual Soldiers and TSUs critical for accomplishing assigned tasks and missions. Additionally, these interfaces must operate in all weather conditions, day and night, without compromising the security of the Soldier or TSU. The same interface doesn't necessarily need to be in use during an entire TSU mission; video may be essential during planning, whereas a single image may suffice during execution. For example, during the planning and rehearsal phase, TSU leaders may want to use large tablet-size devices, but during the execution of a mission, devices should be no larger than a smart phone.

Information Networks

Of most importance is the ability of the TSU to access, understand, and share information. A critical situation being viewed by one member of the TSU should be rapidly and efficiently understood and shared with all members of the TSU. Neuroergonomics might be used in the design of information systems to achieve more efficient operation, especially in minimizing information overload.[6] Potential materiel solutions include the following:

---

[6]As described in the National Research Council report on *Opportunities in Neuroscience for Future Army Applications* (NRC, 2008), neuroergonomics is an emerging field within the broader field of brain-machine interfaces, which explores the ability of the brain to directly control systems beyond traditional human effector systems (hands and voice) by structuring the brain's output as a signal that can be transduced into a control input to an external system (a machine, electronic system, computer, semiautonomous air or ground vehicle, etc.). In the Army context, the goal of neuroergonomics is to facilitate a soldier–system symbiosis that measurably outperforms conventional human-system interfaces (NRC, 2008).

- Adaptive automation. This is a novel neuroergonomic concept for a human-machine system that uses real-time assessment of the operator's workload to make the necessary changes to information systems to enhance Soldier and TSU cognitive performance. This and similar advanced technologies are needed to minimize the detrimental effects of information overload.
- Individual cognitive decision aids. Advanced information systems (including sensors) are needed to provide "actionable" information to a decision maker. Additionally, cognitive decision aids (cognitive agents, decision aids, expert systems, augmented cognition, etc.) would improve a leader's decision making capabilities.
- Position location and tracking information in environments in which GPS signals are strong; but more importantly, those in which GPS-denied (signals are significantly degraded or even blocked) environments (e.g., urban operations). The Army has made great strides in GPS-based systems for dismounted personnel. Significant work is still needed for similar tracking accuracies in GPS-denied environments. Technical areas that show promise are enhanced inertial measurement units, radio frequency ranging/triangulation, and algorithms that manipulate all available information.
- Information being received or transmitted should be tagged to identify who has vetted it, its source, its age, and any other information that allows the users of the information to quickly assess its value to that individual or organization. The tagging and the visualization of tagged information should be automated as much as possible.
- The users (Soldiers and TSU leaders) of information must have the ability to—in an automated fashion—prioritize information for mission command, information collection, and dissemination purposes. For example, during the planning phase of a mission, Soldiers and TSU leaders may be able to handle large amounts of information; however, during the assault Soldiers and TSU leaders should receive only the information critical to the accomplishment of their tasks during that phase of the mission.
- Soldiers and TSU leaders need access to both internal (organic) and external (supporting) sensors for information collection, including those on robotic ground and air platforms. Before deciding what sensor capability is organic to the TSU and what is provided by higher echelons, one first needs to determine which critical capabilities are needed to make a TSU more dominant on the battlefield. For example, the need for real-time remote reconnaissance may be satisfied by a robot, by some other organic asset, or by an asset at a higher echelon. The need may vary during the planning and execution of a mission. Additionally, one needs to ensure that other capability needs (e.g., ability of the operator to develop local situational understanding) are not reduced with the addition of this robot technology or information collection asset.

As a general principle, the design, development, and testing/validation of organic sensor capabilities must ensure that the TSU's sensors contribute to decisive advantage and do not impair other tasks critical to the TSU's operations. For all sensor applications, it is the TCPED process that makes the sensor useful to the warfighter. The TCPED process includes: Tasking the sensor, Collecting and Processing the sensor data, Exploiting information from the data, and timely Dissemination of information to those who need it. In instances of external sensor applications, whole communities are involved with supporting TCPED.

A dismounted TSU does not have the manpower to divert to complicated TCPED activity, yet the TCPED process is needed to provide information relevant to the TSU mission that is timely and actionable. A key challenge for the Army is to figure out how to facilitate TCPED for the TSU while providing the TSU with the degrees of freedom necessary to conduct operations. The effective TCPED solution that will contribute to TSU overmatch is likely to require an unprecedented degree of automation with very low latency. Automation is the only practical way to close the TCPED loop and ensure that organic sensor technology does not adversely preoccupy the TSU's Soldiers. Similarly, the human-system interface is a critical design consideration for organic sensors because anything that affects the unit's cognitive load and ability to focus on immediate task performance requires serious evaluation.

Other considerations on whether an organic sensor capability adds to or detracts from TSU overmatch are the size, weight, and power (SWAP) requirements of the technology. As explained in Appendix G, reducing SWAP requirements is a major factor favoring an open system architecture for sensor technology designed for the dismounted TSU. An open system architecture also provides a foundation to tailor sensor packages for different missions and target types, reducing learning curve and training requirements and simplifying the dissemination of time-critical information.

TSU-organic sensor technology should be developed to meet requirements specifically scaled to the operational needs of the TSU. For example, Appendix G discusses organic situational awareness sensing capability designed to provide a higher level of sensing out to 900 meters from the TSU's location (the primary ring), with a lower level of sensor capability extending to 1,800 meters (the secondary ring). Table G-1 in the appendix and the text accompanying the table expand on these and additional design considerations for TSU-level sensor systems.

Socio-Cognitive Networks

Near-term materiel solutions to integrate the TSU into socio-cognitive networks include:

- Real-time Soldier/TSU access to TiGRNET-like capabilities during dismounted operations and away from static high-bandwidth connections (e.g., hardwire SIPRNet [Secure Internet Protocol Network] connections); and
- Biometric devices built into Soldier/TSU systems to support the recognition of persons of interest during counterinsurgency operations.

In the mid to far term, the goal should be to fully integrate communications, information, and socio-cognitive networks together into a single network. Solutions of particular importance include the following:

- Full integration into the Army network, including integration with autonomous systems networks. Interfaces with autonomous systems would require gesture recognition, in addition to audible and digital interfaces.
- Network-enabled intelligent "Soldier/TSU leader assist" tools to provide alerts of critical information or dangerous situations, assistances with planning and execution of missions, and automatic reporting that requires minimal Soldier/TSU leader input. As an example, the network would quickly begin to populate (with unit location, unit identification, name of individual, current unit activity, etc.) a medical evacuation request once the network detects a critically injured Soldier (keying off sensors that monitor life signs). Then the TSU leader need only add minimal information and hit the send button.
- Network support of information-sharing outside the TSU—for example, sharing information with coalition forces (e.g., operating with host nation forces in counterinsurgency operations)—would require advances in language translation systems and multilevel security systems.
- Soldier devices should be enabled with a full range of biometric sensors. In support of the counterinsurgency operations, the socio-cognitive information network would convey to the Soldier/TSU leader information such as: (1) identification of a person's community, (2) identification of a person's association with overlapping communities, (3) identification of and interaction with local leaders, and (4) the ability to visualize a leader's/person's social connections.
- The network should also identify behavioral trends of both enemy and civilian activities to alert Soldiers and TSU leaders of anomalies.

Integration of the Soldier and TSU into the Army's networks will require near-term investments in Army networks such as the following:

- Communications network enhancements including TSU-level network management, remote control of radio transmission modes, and hands-free display interfaces capable of operating in all weather conditions, day and night, without compromising the security of the Soldier or TSU;
- Information networks capable of providing position location and tracking information in GPS-denied environments, automated tagging of information received to aid visualization, prioritization and dissemination, and access to level 1 situational awareness data from supporting sensors; and
- Socio-cognitive networks capable of providing real-time access to such things as reports on tactical ground activities from collateral units and biometric databases for identification of adversaries.

Network capabilities required in the mid to far term include the following:

- Integration with autonomous systems networks and user interfaces in addition to audible or digital interfaces, such as gesture recognition;
- Network applications, such as an intelligent TSU leader assist tool to provide critical information alerts, assistance with planning and execution of missions,

automatic reporting, and behavior trend analyses of changes in enemy and civilian activities; and
- Network-enabled support of information sharing with collateral forces.

Recommendation 10: To achieve decisive overmatch capabilities, the Army should fully integrate the Soldier and TSU into existing and planned communications, information, and socio-cognitive networks, while ensuring that the network enhancements required for this purpose address all DOTMLPF domains.

Measures (MOPs and MOEs) for assessing levels of situational understanding would have utility for materiel development and evaluation, analytical modeling and simulation, and human factors research, as well as TSU training. It is possible that physiological correlates to such measures could be confirmed, and limited instrumentation could be operational, for validation of materiel development trials conducted, in the mid term. By the far term, it should be possible to assess the range, resolution, and reliability of Soldier and TSU situational understanding in relevant operational environments in real time.

Recommendation 11: In an immediate initiative, the Army should engage the science and technology community(from both human and materiel perspectives), users, trainers, and other stakeholders in Army networks to produce measures for assessing levels of situational understanding needed by the TSU.

BALANCING TSU MANEUVERABILITY, MILITARY EFFECTS, AND SURVIVABILITY

As noted in the introduction to this chapter, the interactive consequences, positive and negative, of any particular capability option can extend across several, if not all, of the five capability categories (situational understanding, military effects, maneuverability, sustainability, and survivability) used in Chapter 2 to describe what dismounted TSUs must be able to achieve across the entire range of military operations. Finding the best combination (or combinations) of options for ensuring decisive overmatch will require balancing these consequences at the system level—for both the TSU system and the Soldier system—as argued in Chapter 3.

A particularly strong level of such interactions occurs for options to improve maneuverability, military effects, and survivability. The committee's initial approach to discussing these three capability categories was to present options for maneuverability, military effects, and survivability separately. However, the draft discussions for these three capability areas kept crossing over into each other. In the context of what the Army expects a dismounted TSU to do—across all the missions and tasks anticipated in future unified land operations—overmatch requires a mission-appropriate balance of maneuverability, survivability, and military effects (including lethal, nonlethal, stability, and humanitarian effects). What the committee found is that, for dismounted operations, this difficult balancing act typically ends up being carried out, literally, on the backs of

Soldiers. For dismounted operations, the fulcrum on which maneuver, survival, and military effects must be balanced is the Soldier's combat load. When the balancing act fails, the consequences degrade TSU and Soldier capability in all three areas.

Soldiers Carry Too Much to Move Quickly, Act Effectively, and Avoid Injury

The Army defines "combat load" as the minimum mission-essential equipment, as determined by the commander, required for Soldiers to accomplish anticipated combat operations. Army Field Manual 21-18, *Foot Marches*, divides combat load into three categories (U.S. Army, 1990):

- *Fighting load*—about 48 pounds of clothing, weapons, helmet, load-bearing equipment, and enough ammunition for the task at hand. However, cross-loading of machine gun ammunition, anti-tank rounds, mortar rounds, and radio equipment will drive load higher than 48 pounds.
- *Approach march load*—about 72 pounds; includes fighting load plus the remainder of basic load of ammunition, small assault pack, lightly loaded rucksack, and poncho roll.
- *Emergency approach march load*—between 120 and 150 pounds; includes approach march load and all other equipment that must be carried when operating in terrain that is impassable to vehicles or when air/ground transportation is not available).

How closely has the Army been able to adhere to the doctrine implied in these definitions of combat load, and how well has that doctrine worked in giving dismounted units decisive overmatch? Based on presentations and discussions with Soldiers, it is obvious to the committee that, in practice, the dismounted Soldier's combat load is far too great, often exceeding the upper limits stated in Army doctrine such as the above definitions. A vignette from recent operations in Afghanistan illustrates how excessive Soldier load can degrade not only maneuverability but also military effects and survivability.

> During Operation Resolute Strike by the 504th Parachute Infantry Regiment, conducted in Afghanistan on 8-9 April 2003, the high desert temperatures, bright sunlight, and approach march loads averaging over 101 pounds per man quickly wore out these physically fit dismounted Soldiers. Each Soldier's water supply was exhausted within the first 12 hours of the operation. The combined effects of the heat and the weight of the combat load made moving even relatively short distances of a few kilometers on relatively flat terrain very debilitating. Unit leaders had to increase rest breaks substantially and drastically increase the resupply of water. Even these trained paratroopers were unable to cope with the physical exhaustion caused by heavy combat loads in harsh climatic conditions.
> (U.S. Army, 2003)

Concern about Soldier loads can be traced back decades. In 1950, Colonel S.L.A. Marshall wrote *The Soldier's Load and the Mobility of a Nation* to address problems with a Soldier's combat load, based on insights and information he collected during the Normandy Invasion in 1944 (Marshall, 1950). Although many changes have occurred in Soldier equipment since World War II, the dismounted Soldier continues to carry his "mission load" on his back, and he is more heavily burdened with mission equipment today than in previous military conflicts.

With such heavy burdens, traversing rough terrain and making rapid changes in direction, speed, and orientation greatly increase Soldiers' susceptibility to injuries. A study by the U.S. Army Medical Research and Materiel Command found that 24 percent of medical evacuations from Operation Iraqi Freedom and Operation Enduring Freedom were due to noncombat musculoskeletal injuries and 72 percent of medical discharges were from chronic musculoskeletal injuries.[7]

As these examples illustrate, excessive Soldier loads degrade not only maneuverability of both individual Soldiers and TSUs but also their resilience, survivability, and effectiveness. Given these wide-ranging negative consequences, why are dismounted Soldiers still carrying excessive load? Among the reasons that TSU leaders mention[8] are (1) weight of the fielded equipment, especially water, batteries, ammunition, and personal armor; (2) mandates from higher-echelon commanders requiring personal armor (individual protective equipment [IPE]) exceeding mission risks; (3) Soldiers' lack of confidence in timely supply (which leads them to want to carry more ammunition, batteries, etc.); (4) doctrinally controlled requirements, such as carrying enough supplies (again, ammunition, batteries, water, food, etc.) for 72 hours of operations; and (5) inadequate delineation of a mission's scope, leading to carrying nonessential items "just in case." An important lesson from this wide-ranging list of probative causes of excessive load is that the load is excessive because the various subsystems and components of the Soldier and TSU systems are being optimized independently of each other. From a systems engineer's perspective, excessive Soldier load and all the capability degradations resulting from it illustrate the suboptimal configuration, to the point of being dysfunctional, of a system (both the Soldier and the dismounted TSU) designed, acquired, and deployed as piece-parts.

The Army approach to addressing excessive Soldier load has been misdirected. The repetition of calls from user and industry representatives for near-revolutionary advances in materials to bring about weight savings to lighten the Soldier load raises unrealistic expectations: Materials weight savings will be at the margins, whereas almost half of the weight is in the bulk items of water, food, fuel (including batteries), and ammunition.[9]

---

[7]COL Gaston P. Bathalon, Commander, Army Research Institute of Environmental Medicine, U.S, Army Medical Research and Materiel Command, "The Soldier as a Decisive Weapon: USAMRMC Soldier Focused Research," presentation to the Board on Army Science and Technology, , February 15, 2011.

[8]The reasons listed here are the committee's distillation from many conversations with squad leaders and platoon and company commanders recently returned from combat deployments, as well as members' reading of news accounts and feature stories from the popular media. The point is not the extent to which one factor or another actually contributes to excessive Soldier load but the *perception* that soldiers are overloaded for what seem to be "good reasons" at the time.

[9]In the report, "The Modern Warrior's Combat Load. Dismounted Operations in Afghanistan April-May 2003, Task Force Devil Coalition Task Force 82, Coalition Joint Task Force 180, OPERATION ENDURING FREEDOM III," water, food, fuel (including batteries) and ammunition accounted for about

Potential Benefits of Optimizing TSU and Soldier Systems for Maneuverability, Military Effects, and Survivability

The above discussion highlighted the negative impacts on TSU and Soldier performance and effectiveness when Soldier load becomes dysfunctional as a result of the imbalance caused by optimizing one or another capability at the expense of others. Just as important for decisive overmatch are the potential benefits of getting that balance right. The following benefits are just a few examples the committee selected for their salience to capabilities for the TSU/Soldier missions and tasks highlighted in Chapter 2.

Situational Understanding

Balancing maneuverability, military effects, and survivability will also enhance situational understanding. For a dismounted Soldier engaged in an operation, it is difficult to concentrate on what is going on around you, let alone interpreting accurately the stream of information coming over communications systems (when fully integrated into the network that is now only fully available at higher echelons) when you are physically exhausted, perspiring profusely, and breathing heavily. Concentration and decision-making abilities, such as required for full situational understanding, suffer as fatigue increases (NRC, 2009).[10] A recent DARPA study on ambushes found that, in the first 5 minutes of an ambush, Soldier and TSU lethality accuracies are quite poor, but increase as the unit settles into the fight.[11] Some of this performance decrement probably results from the response to surprise and disorientation inherent in being ambushed, but the exhaustion caused by quick reaction drills under fire while carrying heavy combat loads also plays a role, as well as constraints on agility. Similarly, threat forces in Afghanistan were known to attack U.S. TSUs just as they were returning from a patrol, when the unit was most fatigued and its situational awareness was degraded.

Military Effects

The physical means for achieving military effects at the dismounted TSU level—personal and crew-served lethal weapons, nonlethal weapons, ammunition, the communications and other electronic devices to call for and guide supporting fires or reinforcements and to conduct stability operations, the power sources these weapons and devices require, and so on—all contribute to Soldier load. Obviously, improvements in the weight efficiency (unit of effect per unit weight) of these carried components and

55 pounds of the emergency approach march load of between 120 and 150 pounds. Water and food accounted for about 33 pounds.

[10]See also the discussion and cited references in the "Physiological Readiness" section of this chapter.

[11]LTC Joseph Hitt, Program Manager, Tactical Technology Office, DARPA, "Lightening the Soldier's Load," presentation to the Committee, December 13, 2011.

subsystems can *in principle* decrease Soldier load or enable other trades that improve overall effectiveness at a given level of load.

Often overlooked, however, are other synergistic enhancements of effectiveness from an optimal balance of maneuverability, the equipment for military effects, and survivability. For example, a significant factor in overmatch is for the dismounted TSU to have advantages in gaining and maintaining surprise or in immediately seizing the initiative even when an opponent acts first, through the ability to outmaneuver the opponent. With high mobility and agility, coupled with superior situational understanding, a TSU can quickly gain and maintain a tactical offensive advantage or initiate protective defensive actions when needed. Coupling enhanced mobility and agility with the right combination of lethal and nonlethal capabilities (including integrated organic and supporting fires) will give dismounted TSUs increased effectiveness in combat operations and the flexibility for appropriate and decisive response across the range of stability operations.

With respect to lethal weaponry, the direct fire capabilities of infantry squads and platoons have improved markedly over the past 10 years through programs under PEO Soldier to improve both personal and crew-served weapons. As discussed in Appendix J, these improvements have reduced the weight of both current weapons and their ammunition, improved reliability, and increased their effective range. Further advances in these areas are currently in development in PEO Soldier programs and in the Army laboratories and engineering centers. A good example of new lethal capability is the XM25 counter defilade system, a shoulder-fired weapon that launches a 25 mm round that explodes at a set distance from the firing point. This developmental weapon, which has been deployed in limited quantities in Afghanistan, gives dismounted units the ability to accurately target enemy combatants behind walls or in other defilade positions that cannot be effectively targeted with other infantry direct fire weapons.

The traditional and most responsive indirect fire system organic to infantry units is the mortar. Mortars have the battlefield role of providing maneuver leaders with immediate indirect area and precision (recently developed) fires. The Infantry Brigade Combat Team (IBCT) is currently the Army's lightest brigade combat team (BCT) and is organized around dismounted infantry. Thus, it provides the clearest illustration of how mortar support is provided to a dismounted TSU under current DOTMLPF. Each of the three types of IBCT (light infantry, air assault, or airborne) has the same basic organization. Within an IBCT, mortars are organic to (found within) each company, battalion, and cavalry squadron, but mortars are not organic to individual rifle squads—the current manifestation of the dismounted TSU on which this report has focused. Infantry battalions serve as the primary maneuver force for the brigade and are organized with a headquarters and headquarters company (HHC), three rifle companies, and a weapons company.[12] Each rifle company has a 60mm mortar section.

The HHC has a mortar platoon with 81 mm and 120 mm mortars.[13, 14] The primary role of this battalion mortar platoon is to provide immediate, responsive indirect fires in

[12]An infantry weapons company has anti-tank weapons (e.g., Javelin) and heavy machine guns (e.g., 50 cal and the MK-19 40mm grenade machine gun).

[13]See FM3-90.6. Available online at https://rdl.train.army.mil/catalog/view/100.ATSC/F8845901-E30D-4488-A923-86825263F32B-1308728592977/3-90.6/chap1.htm.

support of the battalion and its maneuver companies. The battalion mortar platoon provides "general support," with priority to the company involved with the most decisive operation—to reinforce that company's organic mortars.[15] The battalion mortar platoon consists of a mortar platoon headquarters, a mortar section with a fire direction center, and four mortar squads. The platoon's fire direction center controls and directs the platoon's fires. Although each mortar squad in the mortar platoon is equipped with both 120 mm and 81 mm mortars, the authorized size of the squad only permits it to operate one of the two systems at any one time (arms room concept).

Each IBCT rifle company has a mortar section with 60 mm mortars. A company employs these organic (to the company) mortars to support the attack, block ingress/egress routes, and prevent repositioning of enemy reserves. A rifle squad would get supporting fires from the company mortar section or from the battalion's mortar platoon through a call for fire, typically made by a trained forward observer from the company's fire support team. In short, when a rifle squad is operating "on its own" and not as an integral part of a larger company-level action, its access to supporting mortar fire—or any joint fires—depends on having a fire support team member present or having access to the network. As discussed in Appendix J, doctrinal, training, materiel, and leadership changes will be necessary to enable rifle squad leaders to request supporting mortar and other joint fires.

Considering both combat and stability operations, dismounted TSU and Soldier maneuverability and lethality needs vary with roles, missions, and phases of a mission. For example, maneuver TSUs—those that close with and neutralize the enemy—will require more maneuverability than the heavy weapons TSUs in maneuver platoons and especially those in a heavy weapons company. Additionally, since the heavy weapons TSUs are laden with not only heavy weapons (e.g., heavy machine guns, mortars, anti-tank weapons) but also the ammunition for these weapons, their need for improved mobility is more urgent than greater agility. With regard to phases of missions; the TSUs will carry maximum load to assembly areas, a smaller load to pre-assault positions, and finally their combat load during the assault.

Flexibility with respect to military effects becomes even more demanding when TSU mission objectives require a dismounted unit to be prepared to shift rapidly among traditional lethal combat, nonlethal means of projecting force, and stability objectives where effectiveness is measured in terms of communication with the local population, building capacity for civil operations, or humanitarian objectives.

Maneuverability

Tactical maneuverability (combination of mobility and agility) is difficult to achieve in complex, austere, and harsh terrains and at a high OPTEMPO. Mobility for the Soldier and TSU must be equal to or better than adversaries to effectively close with and neutralize the enemy utilizing fire and maneuver. Survivability focused on heavy personal armor will reduce mobility, so survivability ensembles must allow for

---

[14]See FM 3-22.91. Available online at http://www.marines.mil/news/publications/Documents/FM%203-22.91%20Jul%202008%20PT%201.pdf.

[15]See FM 3-21.20. Available online at http://armypubs.army.mil/doctrine/DR_pubs/DR_a/pdf/fm3_21x20.pdf.

adversary-competitive mobility, while keeping casualties within strategic expectations. TSUs also need better maneuverability in complex terrain (e.g., urban, mountainous, jungle). For urban operations, TSUs must not be constrained by ground-level doors and windows for assaulting a building, nor stairwells for vertical movement within a building.

Right-sizing the Soldier's load will obviously enhance both the mobility and agility of the TSU as well as the individual Soldier. But overall TSU maneuverability requires more than just decreasing the Soldier load. For instance, a major concern with the current technology for robot systems that could carry a substantial portion of a dismounted TSU's gear and supplies is whether such systems will be able to "keep up" in difficult terrain or combat conditions and not detract from the unit's ability to maneuver. If the survival of a carrier robot becomes an issue, how much is the unit's mobility and agility in a close fight compromised? Similarly, at what point does scaling down or removing portions of IPE, or leaving back at base a heavy crew-served weapon, make a unit less agile because of increased vulnerability?

Survivability

Solutions to assure survivability at the Soldier and TSU level range from developing superior weapons and sensors to technologies for ballistic and climate protection to fundamental considerations for lightening the Soldier's load. Across the range of military operations, the protection function alone consists of "...capabilities to identify, prevent, and mitigate threats to assets, forces, partners, and civilian populations to preserve combat power and freedom of action." (TRADOC, 2010)

As a result of Iraq and Afghanistan, many people would believe that the infantry Soldier's survivability in the future depends only on better and lighter armor protection.[16] This emphasis on increasing the ballistic protection of the Soldier to increase survivability has hindered maneuverability and endurance of both the dismounted TSU and the individual dismounted Soldier by adding to Soldier load and constraining Soldier agility.

For dismounted missions, those capability decrements can have substantial negative effects on survivability and military effects. Although optimization studies of the trade between increased Soldier ballistic protection and degraded capability to maneuver, endure, and act effectively may have been done, the committee could find no evidence of such trade studies. If the survivability benefit of increased IPE weight were optimized against the weight and agility-constraint consequences, the committee believes that survivability can be significantly enhanced through such indirect consequences as the Soldier's and TSU's abilities to maneuver more effectively against an enemy and maintain or seize the offensive. The hard part is finding the right balance of IPE with other factors that contribute to Soldier load. In addition, such analytical exercises are only of real value to ensuring future overmatch when they are built on realistic, validated measures of performance and effectiveness.

---

[16]Soldier and TSU protection includes not only personal and vehicle armor but also operating base protection and protection during movement.

An optimal balance of maneuverability, survivability, and means for military effects also has beneficial second-and third-order effects on survivability. For instance, decreasing Soldier load may reduce heat exhaustion as well as physical injuries—especially musculoskeletal injuries. Without the degradations to performance and effectiveness from these environmental injuries, dismounted TSUs and Soldiers will be less vulnerable to combat-related wounds and death.

Selected DOTMLPF Opportunities for Balancing Maneuverability, Military Effects, and Survivability

Opportunities for balancing maneuver, survival, and military effects fall in all of the DOTMLPF categories. The *optimal* balance at the system level is unlikely to be simply a matter of reducing Soldier load, improving weapons and ammo, using a robot carrier, or any other single, materiel-focused approach. As argued in Chapter 3 and again in the "Designing the TSU" section of this chapter, there are multiple options for improving capability in one area or another, but ensuring decisive overmatch requires putting together the whole package and, most important, ensuring that a contribution in one capability is not outweighed by unintended decrements to other capabilities essential for overmatch across the entire range of dismounted TSU missions and tasks.

The entire panoply of potential opportunities cannot be explored here. The committee has selected a few examples that: (1) seem to have the highest potential payoff, from the limited knowledge base available to the committee, (2) illustrate the importance of considering opportunities (and negative impacts) across the full range of DOTMLPF, and (3) can be addressed meaningfully in the near, mid, or far terms (within 5 years, 5-10 years, and beyond 10 years, respectively).

Doctrine

For reasons of resupply, the Soldier's combat load is currently based on mission durations of 48 to 72 hours. Mission durations are decided upon by unit leaders based on experience and mission needs, but the 48 to 72 hour duration also reflects the guidance provided in doctrinal documentation and unit standard operating procedures. A substantial fraction of a Soldier's load is the basic load of food, water, ammunition, batteries, etc., required for 48 to 72 hours of operations. This basic load requirement drives up the weight of the approach march load (as in the example of Operation Resolute Strike, described above) and especially the emergency approach march loads These loads are further increased for Soldiers carrying heavy weapons and the ammunition for them.

Given this dependence of the basic load on mission duration, doctrinal guidance on mission duration needs to be evaluated in light of the experience gained with the performance and effectiveness consequences of current Soldier loads in challenging environments during operations in Iraq and Afghanistan. Advances in load-carrying technologies (e.g., semi-autonomous logistics robots), resupply and sustainment technologies (e.g., renewable energy systems for recharging batteries), access to automated reports in network technologies, integration of fires (which might reduce dismounted TSU loads for heavy weapons ammunition), and other advances in materiel technologies need to be evaluated with mission duration as an independent variable, rather than assuming one nominal mission duration.

TTPs also need to be established for the optimal utilization of mobility planning tools, load-carrying technologies (e.g., robotic platforms), enhanced logistics capabilities (e.g., aerial resupply, use of renewable energy systems), and changes (if any) to the TSU organization, as described below.

Force protection countermeasures to date have largely relied on IPE that has added weight to the Soldier's load, reduced Soldier and TSU maneuverability, and because the Soldiers are less agile, made them more vulnerable to enemy fires. Other doctrinal approaches to force protection that would help reduce Soldier combat loads need to be considered. For example, the force protection benefits of integrating the TSU into the Army network—especially for improved integration of supporting fires—should be evaluated. Such analysis may demonstrate that improved integration of fires reduces the required amount of organic fires, thus reducing the amount of heavy weapon munitions to be carried within the maneuver TSU and supporting-weapons TSUs.

Organization

Very old studies found that, as the size of a squad decreases, its maneuver becomes more successful (Marshall, 1950). However, analogous studies need to be conducted today to determine if the same findings are supported, in the context of current doctrine, mission command technologies, and training initiatives, for operational scenarios characteristic of what the Army expects dismounted TSUs to do in the future.

Such studies might consider, for example, the relative performance and effectiveness of smaller, more agile fire teams within the TSU. In the Army's dismounted squads today, the two fire teams within a squad are organized the same way with respect to their size, weapons, etc. Would a squad with two "light" fire teams (three or four Soldiers each), carrying only rifles, plus a third fire team of four Soldiers carrying only heavier weapons (e.g., M249 squad automatic weapon, M203 machine gun, and XM-25 counter defilade engagement system) be more effective, and in which scenarios?[17] A TSU organized this way would allow its two very mobile and agile fire teams to conduct swift maneuver operations while the third heavy-weapons team provides covering fire. As with other suggestions for optimizing the balance of maneuver, survival, and means for military effects, the point is not to rely on opinions for or against the current organization or this suggested alternative but to conduct controlled experiments to provide the data necessary for valid trade studies.

[17]Appendix H reviews the individual and crew-served weapons currently fielded, in development, and projected for future development. The M249 and M203 are currently fielded, the XM25 is considered in development, although it has been used operationally, as noted in the appendix.

Training

The section above on "Focusing on TSU Training" dealt primarily with the training needed for all Soldiers in a TSU to optimize their performance as a unit. With respect specifically to the balance issues addressed in this section, all Soldiers need to be sufficiently physically fit to carry combat loads—even after those loads have been significantly reduced from recent excessive levels—for extended distances over all terrain and in all weather conditions. More-focused training programs will be needed for unique mission needs, such as training for operations at high elevations. Training programs need to take into account the physical capabilities of the personnel volunteering for military service; the combat veterans with whom the committee talked reiterated many times the point that most Soldiers entering basic training could not pass the physical training program.[18]

With respect to achieving and sustaining the optimal balance of maneuver, survivability, and military effects in operations, training for TSU leaders should include instruction on factors that affect squad mobility, including terrain, meteorological conditions, loads, load configurations, accumulated fatigue, IPE, and how factors like IPE and load configuration constrain agility. Leader trainees should be given practical exercises to increase their confidence in the validity of their planning aids. Soldier load planning and the mobility and endurance effects of different loads should be factors in all training simulations and games.

Materiel

The materiel opportunities for optimizing the balance of maneuver, survivability, and military effects are quite varied in both the capabilities they enhance and the potential decrements they may cause. They should be assessed in an integrated evaluation process that takes into account the non-materiel components of DOTMLPF as well as the capability interactions among TSU/Soldier materiel components and systems. The examples discussed here, selected to illustrate the variety of opportunities, are load-carrying systems designed for use by a dismounted TSU, exoskeletons, lethal/nonlethal weaponry, IPE, mobility planning aids, improved resupply for TSUs engaged in an extended operation, Blue Force tracking technology at the individual Soldier level, and rations.

Load-carrying Robot Systems. Past attempts to offload the Soldier's logistics burden to a manned or unmanned carrier have not been successful. However, a carrier, either manned or unmanned, might improve TSU tactical maneuverability by providing such things as information collection (formerly called "ISR"), battery recharging, or casualty transport. TSU load-carrying systems, such as the Squad Mission Support System and other semi-autonomous and autonomous systems described in Appendix H, should be

[18]Group discussion between the committee and a group of Army noncommissioned officers recently returned from deployment, U.S. Army Maneuver Center of Excellence, Fort Benning, Georgia, July 12-14, 2011.

considered for reducing the combat load—especially the approach march load—of the Soldier. Soldier and TSU maneuverability is hindered most by combat loads in very complex terrain and harsh weather conditions. These load-carrying systems need to be designed to operate in these same challenging conditions.

Exoskeletons. Exoskeleton suits have come a long way in the past several decades. They seem less bulky, more responsive, and more aligned to warfighter needs (as currently perceived)—albeit for a limited number of those needs. Some issues of immediate concern for exoskeleton use by dismounted TSUs are their impact on agility and battery power requirements, as well as their hydraulic actuator systems, sensors, durability, maintainability, and reliability. Even with these issues not fully resolved, there may be a near-term use for exoskeletons in infantry heavy weapons units (weapons TSUs in a platoon and the heavy weapons platoon in a company). These units carry medium machine guns (7.62 mm M240), anti-tank systems (Javelins), and ammunition for both systems. The amount of carried ammunition limits sustained engagements. Therefore, the more ammunition that can be carried by a dismounted TSU, the more lethal that unit can be. Also, a heavy weapons TSU does not need to be as agile as the maneuver (or line) TSU, since it is usually deployed in overwatch positions and not as the first unit to make contact with the enemy. Similarly, these systems may be useful for heavy weapons platoons: those carrying the heavy 0.50 caliber machine guns, mortars, and associated ammunition. In the far term, more advanced generations of exoskeletons may offer benefits to maneuver TSUs that outweigh the negative consequences of current technology.

Lethal/Nonlethal Weaponry. As discussed in Appendix J, the Army has multiple ongoing activities aimed at improving the individual weapons available to dismounted Soldiers, the crew-served weapons that a dismounted heavy weapons unit might use, and an expanding array of nonlethal weapons for use in combat or combat-related stability operations. From an operational perspective, it would be beneficial to have multi-mode weapons that allow users to easily switch from lethal to nonlethal mode and back to lethal without requiring the Soldier to physically switch weapons. TSU leaders and Soldiers need to be trained not only for proficient use of each lethal/nonlethal option but also on the TTPs to guide which options they employ under which circumstances. However, the development of nonlethal options must be informed by a better understanding of the behaviors to be expected from targets threatened with or engaged by a particular nonlethal weapon. In both combat and stability operations, if those being confronted perceive no differences in the visual and acoustic signatures of brandished lethal and nonlethal weapons, their response will likely be to presume the weapon is lethal. Without good understanding of the expected behaviors of the intended targets for nonlethal effects, a dismounted TSU's shift to nonlethal effects could lead to an unexpected escalation from those confronted.

Additionally, given the potential threat of improvised remote control drones (even toy drones), consideration should be given to the development of TSU-level counter-improvised-drone weapons (e.g., an improvised equivalent of the XM25) and munitions (e.g., a 40 mm round that fires a wide dispersion of buckshot).

IPE for Dismounted Soldiers. Key opportunities with respect to IPE include reducing weight and stiffness while increasing body cooling. IPE is very heavy (above recommended weights) and often very stiff (e.g., plates in IPE). The combination detrimentally affects both mobility and agility. The IPE, in combination with other carried equipment and kit, inhibits natural cooling of the Soldier's body. Weight and bulk of IPE must be reduced while making the design of IPE much more flexible and comfortable to wear. Replacing hard protective plates with flexible, lightweight systems that conform to the body, minimally impede motion, and permit body cooling would be a major improvement. The committee also heard that the Army continues to have difficulty with properly fitting Soldiers with clothing and kit, including properly fitting IPE.[19]

A significant research, development, experimentation, and demonstration program could be initiated to integrate protective equipment with passive as well as active cooling technologies. If up to 60 percent of the physiological load from tactical load carriage derives from the need to dissipate heat from the near encapsulation of the body core with armor, while alternative cooling technologies have had some limited demonstration but little serious R&D, there appears to be substantial potential for improving this aspect of Soldier load by employing passive cooling technologies. Cooling must be seen as directly enabling full integration of Soldier-worn technologies. That is, if Soldiers find the integrated ensemble unbearable on long missions, unit survivability is compromised. If full integration is the first goal, cooling will be relegated to a lower priority as a secondary accessory. The criteria for evaluating weight saving integration should be validated TSU metrics (MOPs and MOEs) including the propensity for chronic injuries.

Materiel developers offered that IPE development and manufacturing programs go to great lengths to ensure sufficient sizes are available to effectively fit the diversity of body shapes and sizes in the Soldier population. However, they noted their surveys showed that a significant portion of solders in the field have been issued the wrong size, usually degrading their mobility to a very measurable degree.

Mobility Planning Aids. TSU leaders need a "mobility planning aid" that would predict TSU mobility in terms of speeds for both endurance distances and rushing sprints, as a function of (1) terrain (specific routes including elevation), (2) meteorological factors (temperature, humidity, wind, solar loading), (3) ration intake and hydration, (4) loads (including IPE), (5) physical attributes of the individual TSU members (fitness, anthropometry, degree of misfit of IPE), and (5) resupply points. Such a planning aid could identify the unit-specific member-by-load combination most likely to be the limiting factor in the unit's mobility. The empirical basis for this planning aid should be developed to also predict the incremental risk to long-term injury the mission will contribute to each TSU member. Critical to getting this type of planning aid "done right" is putting together an integrated team that includes the relevant centers of expertise in Army organizations. For example, this effort might be led by PEO Soldier but should include the Maneuver Center of Excellence Infantry School, USARIEM, PEO STRI, the Natick Research, Development and Engineering Centr, and ARL/HRED. However the

[19]Based on committee interviews with R&D personnel at the Natick Soldier Research, Development and Engineering Center, Natick, Massachusetts, September 15, 2011.

Army decides to lead and staff such an initiative, it must adhere to the principles laid out in Chapter 3.

Improved Resupply. Enhanced, reliable resupply systems need to be developed so as to reduce approach march loads and especially emergency approach march loads. These systems include, but should not be limited to, autonomous ground and air systems, precision aerial delivery systems (e.g., GPS-guided parachutes), and foraging/harvesting systems (e.g., renewable energy systems).

Blue Force Tracking Technology for Individual Dismounts in the TSU. TSU maneuverability could be enhanced if small unit leaders had immediate knowledge of the locations of subordinate fire teams and individual Soldiers, especially in night operations, obscuring environments, or complex terrain (e.g., urban structures). Network integration technology, such as Blue Force unit tracking in complex terrain, should be adapted to provide this Soldier-level enhancement to support TSU maneuverability.

Rations. Over the past several decades, the Army has done tremendous work in improving combat rations for both in-base and combat-patrol consumption. Work should continue to reduce the bulk and weight for a given amount of nutrition without sacrificing palatability.

Leadership

TSU leader training should include instruction on factors affecting unit mobility as well as on the uses and benefits of maneuver-supporting materiel subsystems and components (planning aids, load-carrying robot platforms, renewable energy systems, aerial resupply, etc.) For instance, all TSU leaders need to be aware of how the nutritional, medical, and physical training needs of their personnel affect unit performance. Leaders need to have access to appropriate support to meet these needs.

Personnel

More intelligent approaches are needed to the initial selection of personnel for the Infantry branch and then later for TSU positions. TSUs, especially those already deployed or close to deployment, cannot—without sacrificing performance and effectiveness—take on Soldiers who are not physically capable of the demands of dismounted TSU operations, especially in complex terrain and harsh weather conditions. Similarly, the physical capabilities of the individual must be considered when assigning TSU weapons to Soldiers. For example, very agile, 120-pound Soldiers should be considered for rifleman positions as opposed to assignments as light/heavy machine gun, anti-tank weapon, or mortar personnel.

Mental agility (the ability to think and draw conclusions quickly; intellectual acuity) is extremely important to the overall maneuverability, military effectiveness, and survivability of the TSU and its Soldiers. As noted in chapter 2, mental agility is not subsumed under maneuverability. Rather, it is a critical Soldier and TSU capability that can be supported and extended by Level 3 situational awareness and in turn supports decision-making for use of military effects. If the goal is TSUs with decisive overmatch, then mental agility must be a criterion for TSU leadership positions.

Finally, from a personnel perspective the musculoskeletal injuries problem must be addressed within all relevant DOTMLPF domains. Efforts must be made to address this very debilitating problem, which is diminishing the readiness of Army forces today while ballooning the future, long-term medical needs of Army veterans.

Facilities

To maintain adequate physical conditioning, Soldiers should have access to weight training and aerobic exercise equipment in base camps. Soldiers also need access to facilities that help them meet nutritional and health needs. Finally, facilities are needed to support simulation-based maneuver training, even while in combat base camps. These same simulation-based training capabilities can support TSU mission rehearsals. (See related discussion of training technologies in the "Focusing on TSU Training" section in this chapter.)

Findings and Recommendations for Achieving TSU Balance

Assessing Alternatives for Balance in TSU Maneuverability, Military Effects, and Survivability

A priority consistent with the Statement of Task for this study is assessment of both the degraded mobility and the often chronic injuries caused by the heavy loads carried by dismounted small unit Soldiers. Senior Army leaders commented to the committee that loads of 100 pounds or more are excessive, even while acknowledging that such warrior loads can be traced to Roman times. Small unit leaders attested to the debilitating effects of the excessive loads and attributed the problem to a variety of factors, as detailed above in the section on the Soldier load problem.

Finding: Current alternatives offered by the technology communities for addressing Soldier load are to lighten the items carried and to offload sustainment materials to field robotic vehicles. Innovative concepts (as well as refinements of existing capabilities), operations research evaluations of those concepts, and field trials and demonstrations of those concepts are needed to determine which options are more promising. Airlift with precision airdrops; small (non-robotic) 4x4 vehicles; and changes in operational tactics to allow daily resupply should all be evaluated. The focus of this effort should be on attaining operational solutions rather than technology sophistication, and the impact on the Soldier load must be considered explicitly. All DOTMLPF domains need to be considered in all simulation-based and field trial evaluations. Criteria for evaluation of alternatives should employ the TSU metrics (MOPs and MOEs) discussed in Chapter 3, and these metrics should be adequate to assess contributions and factors from all

DOTMLPF domains, with particular emphasis on probability of reducing both acute and chronic musculoskeletal injuries.

Finding: Improvements in fitting skill and knowledge as well as in the distribution of issued individual protective equipment offer potential for improving the mobility of individual Soldiers. Integration of individual protective equipment with passive and active cooling technologies offers potential to improve Soldier performance.

Finding: Experimental trials are needed to develop models for predicting the vulnerability of dismounted individual Soldiers and TSUs as a function of Soldier load and measures/indicators of individual/TSU mobility and agility such as dash speed (e.g., cover to cover). Combat engagement factors included in these trials should include visual detection, identification, and targeting of the opposing element for relevant combat-encounter scenarios (e.g., Blue Force-initiated contact, ambush of Blue TSU, urban/village setting with sudden transition from stability operation to lethal fight). Environmental factors including terrain, elevation, and weather would be later parameters to add to the models and scenarios incorporated in the trials. TSU mobility models must intimately interact with task-workload models to be used to assess information collection and weapons technologies offered as candidate equipment for TSUs. The dismounted Soldier mobility models currently in use at the Natick Research, Development and Engineering Center appear to have value as a starting point for developing models that meet the requirements for realistic and validated evaluation of both current alternatives for addressing Soldier load and innovative concepts. One approach to implementing the experimental trials and model improvements envisioned here could be through a consortium involving ARL/HRED, USARIEM, the Army Test and Evaluation Command, the Army Materiel Systems Analysis Activity, and the Maneuver Center of Excellence Infantry School. The objective is to bring together expertise from across the currently stovepiped and dispersed centers of relevant expertise under the oversight and direction of a high-level systems engineering entity consonant with the principles set out in Chapter 3.

The types of engagements included in these trials need to cover the range of engagement scenarios that dismounted units may encounter in future unified land operations, including stability tasks as well as combat encounters. The goal should be to enable development of realistic, validated models for use in evaluating a wide range of current approaches and innovative concepts for managing Soldier load to achieve an optimal balance of TSU and Soldier maneuverability, military effects, and survivability.

Recommendation 12: The Army should initiate and maintain a program of experimental trials to inform improved models for assessing the effectiveness of dismounted Soldiers and TSUs as a function of Soldier load and measures/indicators of mobility and agility. The program should include an iterative process to explore innovative concepts for balancing TSU maneuverability, military effects, and survivability, as well as continuing exploration of more traditional approaches such as lightening individual items carried and offloading Soldier load onto robotic carriers.

Nonlethal Alternatives

Finding: The full range of prospective operations for dismounted Soldiers and TSU is very likely to exceed that experienced in current conflicts. Stability operations especially will require a mix of lethal and nonlethal capabilities that are not currently available. The emphasis in the nonlethal weapons R&D community appears to be on expanding the menu of options available to warfighters, with the expected outcomes from use of nonlethal effects being as straightforward as for lethal effects. However, nonlethal weapons should be used with an expectation of initiating a specific behavior on the part of the targets. There appears to be little research on understanding the behaviors to be anticipated with each of the nonlethal weapon technologies and the variance of these behaviors among cultures. Also, there appears to be little understanding of the engagement decision complexity that could come to individual TSU Soldiers, with attendant lengthened decision cycle times and greater opportunities for errors.

For managing Soldier load and simplifying dismounted kit, one or more "weapons" that can be readily shifted between lethal and nonlethal modes would be useful. But two downsides to such multimodal weapons are that (1) Soldiers must be well trained on the rules of engagement (ROE) and TTPs for selecting the right mode for a given situation, and (2) in noncombat encounters such as stability operations, signaling to the other side in an encounter that a multimodal weapon is in nonlethal mode could be mission-critical. The effectiveness of *nonlethal* actions, as an alternative to *lethal* effects, will depend to a great extent on the perceptions of those being confronted.

Recommendation 13: In the mid-term, the Army should undertake research to identify a range of unambiguous signals of nonlethal intent. The research should extend to the exploration of cultural differences in intent interpretation.

TSU Mission Planning Aid

Finding: TSU leaders and their commanders at higher echelons need to understand how factors across all the DOTMLPF domains affect not only Soldier load but also the more encompassing goal of balancing maneuverability, effective action, and survivability to ensure small units have decisive overmatch wherever and under whatever circumstances they operate.

Given the range of missions and tasks that dismounted TSUs may be called upon to perform in the future, even experienced leaders at the TSU level and higher echelons cannot be expected to know immediately the best combination of available options, extending across all DOTMLPF domains, for the optimal balance of maneuverability, military effects, and survivability in every environment and engagement. An easy to use mission planning aid could incorporate the relationships among options learned from prior operational experience (lessons learned), as well as the relationships among metrics, indicators, and DOTMLPF options found and validated through experimental trials and incorporated in assessment models used by the development community.

Properly designed, such a mission planning aid would include long distance endurance and sprint speed (as surrogates for engagement vulnerability), functions of terrain (specific terrain route, including elevation), meteorological factors (temperature, humidity, and wind), ration intake, loads (including IPE), physical attributes of TSU members (fitness, anthropometry, degree of misfit of IPE), and resupply points. It would identify the TSU member-by-load combination most likely to be the mobility limit for that particular TSU. If the empirical basis could be developed, the planning aid could also predict the probability that the mission would contribute to the long-term injury or disability of particular TSU members.

The mission planning aid would be used in training TSU leaders on the factors that affect squad mobility, including terrain, meteorological conditions, loads, load configurations, accumulated fatigue, IPE, and how factors like IPE fit and load configuration constrain agility. Practical exercises for leader trainees would increase confidence in using the planning aids in operations. Also, the aids to Soldier load planning and mobility and the endurance effects of different loads could be incorporated in training simulations and games.

Recommendation 14: The Army should develop a mission planning aid to assist in balancing maneuverability, military effects, and survivability, for use in training and operations by TSU leaders and leaders at higher echelons.

LEVERAGING ADVANCES IN PORTABLE POWER

The importance of overcoming limitations placed on Soldier and TSU operations related to ensuring adequate power sources cannot be overestimated. The last decade has seen major advances in portable power materiel technologies, which could have outsize influence on overmatch. However, this can occur only if the Army can leverage the advances to their full effect.

DOTMLPF Considerations

Portable power issues have doctrinal implications because of their impact on TSU TTPs. Non-rechargeable (primary) batteries tend to be less expensive to purchase, have greater energy storage density, have longer shelf life, and, in general, are safer than rechargeable batteries, which may become unstable or even hazardous in extreme temperatures or charge states. Rechargeable batteries also have a relatively limited life and must be kept charged to avoid deterioration. For these reasons, primary batteries have been the predominant type used in operations, while rechargeable batteries have been used primarily for training, with only limited use in operational environments. Any portable power solution that incorporates battery recharging as a key element must therefore address these reasons why rechargeables have not been widely accepted in operational TTPs.

There are also personnel and leadership considerations. To adequately prepare for missions, Soldiers must have an accurate accounting of the state of all their equipment—weapons, bullets, equipment, water, food, and energy provisions (including batteries). To better understand battery status, advanced “gauges" are needed to give trustworthy estimates of remaining capacity. For example, even without use rechargeable batteries lose storage capacity over time, and a 100 percent reading of the charge level is a false representation of the originally designed capacity—it may well be only a fraction of the original energy storage capacity.

The size and weight of power sources can also dictate the duration of operations by affecting the load that must be carried on the operation or the particular electronics that can be utilized. These considerations in turn can affect the optimal organization of the TSU as well as training, leadership, personnel, and facility requirements.

Figure 4-2 lists prospective Soldier power solutions to meet the Soldier energy demands as compiled by the Army in the near, mid, and far terms. Appendix I discusses the state of the art in the underlying Soldier energy technologies, including the battery, fueled, and energy-harvesting systems that are listed in Figure 4-2. While the range of solutions is definitely impressive, each entry on the list comes with its own set of DOTMLPF challenges that must be met to make measureable improvements in dismounted Soldier and TSU capabilities.

Battery and Fueled Energy Storage Systems

Considering the current squad organization and equipment, batteries remain the energy source of choice for missions less than 72 hours. Batteries range in sizes from button cells to large single cells that are arrayed in series and parallel to achieve the requisite energy storage, pack voltage, and acceptable discharge rate for the variety of equipment required. However, Soldier criticism of battery technology is very specific and has formed the basis of the Army R&D program. Materiel shortcomings of batteries include:

- Too many battery types;
- Not energetic enough;
- Too many batteries needed for long missions;
- Too heavy and bulky; and
- Evolution of capabilities adds to energy requirements.

As described in Appendix I, battery systems will continue as the mainstay energy source for the Soldier either as a stand-alone source or as a component of an air breathing hybrid configuration. The specific energy of rechargeable batteries is approaching the specific energy of today’s primary batteries, and advances in rechargeable lithium-air batteries now provide battery-like performance on a par with fuel cells.

Soldier Power Solutions

	2014	2024	2030+
Energy Storage (Disposable)	• Replace Li/SO2 (1X) with Li/MnO2 (1.5X) • Introduce Li/CFx (2X) • Charge indicators – "fuel gauge on each battery"	• Replace Li/MnO2 (1.5X) with Li/CFx (2X) • Introduce Li/Air (4X) • Improved battery chemistries: Li/CFx, Li/MnO2, Zn/Air, Li/FeS2 and Li/Air • Integrated Hybrid battery/ultra-capacitors	• Replace Li/CFx with high performance Li/fluorocarbon • Replace Zn/Air with Li/Air • Bio-inspired materials • Carbon nanotubes • Mini and micro batteries
Energy Storage (Rechargeable)	• Nano-Li-Ion • Li-Ion Polymers • Vehicle based large scale recharging stations • Smart batteries • Battery health and energy management displays • Conformal shaped batteries (140wh/kg)	• Improvements in nano-technologies (e.g. Nano-Li-Ion and LiTitanate) • High voltage cathode materials • Wide window electrolytes • Integrated Hybrid battery / ultra-caps • Wireless recharging • LiMetalF chemistries	• "True" Li-Polymers (flexible, conformal solids) • Bio-inspired materials • High temperature electrodes • Integration into body armor • Molten salt electrolytes
Energy Harvesting	• Thin film photovoltaics (backpacks, foldable tail/wing) • Motion capture, e.g. hand cranks, knee braces, backpacks, Micro-inverters	• Photovoltaic textiles • Motion and vibration devices • Thermoelectrics from body heat • Inductive charging	• Thermoelectrics, combining heating and cooling • Backpack swing generators
Fueled Systems	• Fuel cells (25-75 watts in rucksacks, 150-300 watt and 25-35 lbs man-transportable); packaged fuels (1300 wh/kg) • Engines (portable, multi-fuel burning Stirling 45 lbs) • Small diesel engine 35 lbs	• Fuel cells (25 watts on Soldier, 500 watt 20 lbs man-portable); desulfurized JP8 • Thermoelectric/thermo-photovoltaics (50-150 watts, 25 lbs, 15% conversion efficiency, <70dba)	• Fuel cells (standard JP8) • Carbon nanotubes • Thermoelectric and heat scavenging • Thermophotovoltaics • Multifuel, cheap, reliable engines

Prepared by: Army Capabilities Integration Center – Research, Development and Engineering Command – Deputy Chief of Staff, G-4, US Army (April 2010)

FIGURE 4-2. Soldier power solutions. SOURCE: U.S. Army, 2010.

Finding: Rechargeable lithium-air energy sources used as the primary energy source in hybrid configurations can replace many primary and rechargeable storage systems now in use.

In addition to small fueled engines, the Army has focused on developing several types of fuel cells for a wide range of applications ranging from "wearable" energy sources to large battery chargers. Small fuel cells applicable at the Soldier and TSU level are sufficiently advanced that they are being evaluated in the field. An advantage of fuel cells is that they have low acoustic and thermal signatures. A major drawback to current fuel cells is that they cannot operate on JP (the battlefield logistics) fuel.

Figure 4-3 illustrates the potential of the various energy storage options and provides a basis for committee findings on technology solutions considered for the TSU. The figure depicts the mass of current systems needed to provide operational kilowatt-hours of energy. Systems illustrated for comparison include six primary or rechargeable batteries (standard inventory lithium BA5590 and BA5390, lithium-polymer, lithium-air, and lithium-carbon monofluoride) plus two fuel-cell systems (direct menthanol and solid oxide). Detailed characteristics of these energy systems are discussed in Appendix I.

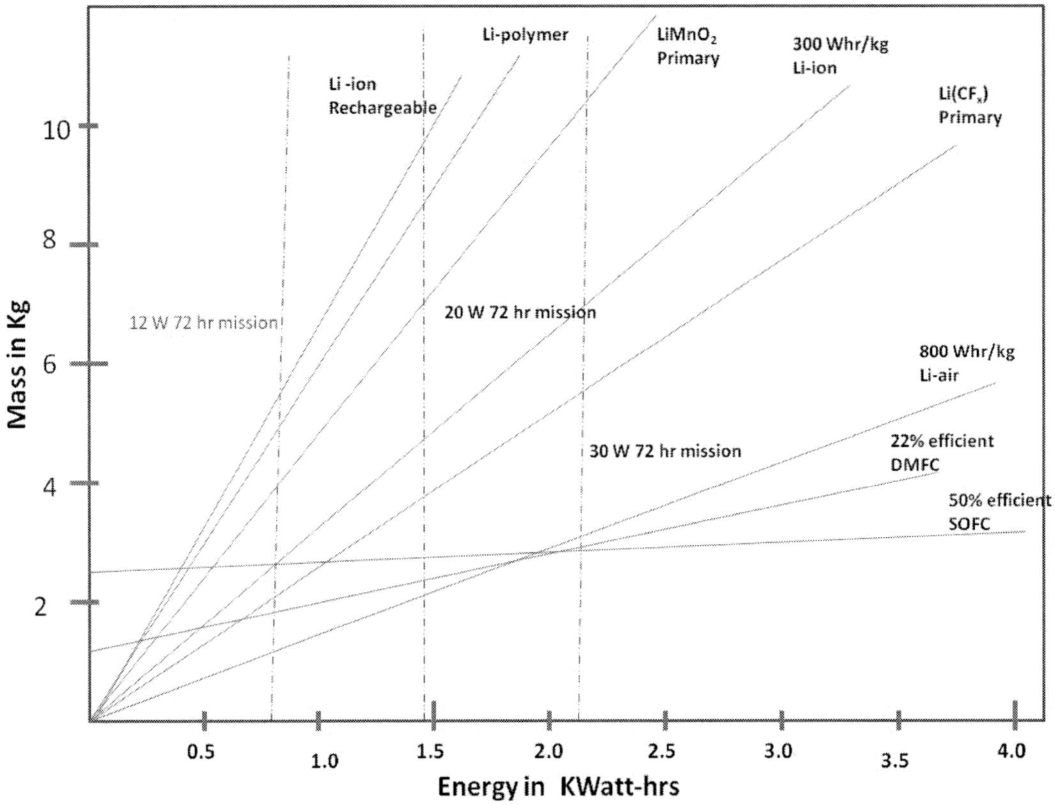

FIGURE 4-3. Comparison of energy options for the dismounted Soldier. SOURCE: Adapted from NRC, 1997.

The selection of a rechargeable battery storage technology as the principal choice for the dismounted Soldier's energy source would necessitate the parallel introduction of a recharger technology sufficiently small and lightweight to be applicable at the dismounted TSU level. JP-fueled motor generators for possible use in rechargers do not scale favorably to small sizes. The successful development of a JP-fuel reforming technology would allow for small combustion engine battery chargers of low cost and light weight.

JP fuel has a weight advantage with respect to carrying additional batteries, given that it has about 10 times the available energy on a per-kilogram basis. But any calculation of the tradeoffs would need to include the weights of the JP container as well as the fuel-cell energy converter. A concept of operations for such a promising opportunity would have to weigh all factors and consequences,

Finding: JP-reforming technology will have to be developed over a wide range of sizes before the Army can exploit either rechargeable battery technology or fuel-cell technology.

The most important development for the dismounted Soldier in the near term is a rechargeable battery-based conformal central power supply to power the Soldier's equipment ensemble. Integrating a single power source would standardize connectors and enable the Army to take maximum advantage of the best and lightest of whatever technology is available—weapons, navigation, communications, pointing devices, etc. For the mid-term, a Soldier-portable battery recharger, powered by a logistics fuel available in theater, would dramatically extend mission duration. The most important far-term development is a rechargeable metal-air energy system (with specific energy approaching that of fueled systems).

While the Army is well on the way toward developing a rechargeable battery technology to become the principal energy source for the Soldier on the battlefield, aside from the materiel development itself, critical DOTMLPF elements have not been evaluated.

Finding: There is no doctrinal philosophy for the tactical small unit to recharge the battery: there is no organizational equipment to support recharging; there is no hint of what training would be required; and, there is no parallel materiel development of the recharger or fuel reformer that would be needed.

Energy Harvesting

Solar and biomechanical energy harvesting systems have been developed to the point where evaluation by Soldiers is possible. Solar battery chargers are in the inventory. Use of photovoltaic cells with higher conversion efficiencies will reduce the weight and volume of solar harvesting techniques for use as battery chargers. Use of biomechanical harvesting will increase as the demand for energy by the Soldier decreases due to increases in the efficiency of Soldier equipment.

If the Army can somehow reduce the energy demands of Soldier equipment, energy harvesting could be used to provide a significant amount of the individual Soldier's energy requirement. Schemes for harvesting energy must generally be used in hybrid configurations. Of the several harvested energy systems discussed in Appendix I, the most relevant at the TSU level for the near and mid terms are portable solar systems and biomechanical systems that extract energy from individual-Soldier movement.

Finding: The full impact of energy harvesting mechanisms on Soldier and tactical small unit performance has not been determined.

Leveraging these advances in energy sources will help to reduce Soldier fatigue, eliminate Soldier anxiety associated with tenuous resupply, increase Soldier confidence in situational awareness from powered sensors, and provide needed assurance that communications links with higher levels in the command structure can be maintained.

Finding: Portable power advances can best contribute to the decisiveness of future Soldiers by increasing the certainty of Soldiers that their equipment ensemble will have sufficient energy to carry out any TSU mission.

Recommendation 15: The Army should develop and maintain a robust program in advanced energy sources based on full analysis of DOTMLPF elements, with the goal of eliminating power and energy as limiting factors in TSU operations.

REFERENCES

Alliger, G.M., R.L. Beard, W. Bennett, Jr., and C.M. Colegrove. 2012. Understanding Mission Essential Competencies as a Job Analysis Method. Pp. 603-624 in The Handbook of Work Analysis: The Methods, Systems, Applications, and Science of Work Measurement in Organizations, edited by M.A. Wilson, W. Bennett,Jr., S.G. Gibson, and G.M. Alliger. Florence, Ky.: Routledge Academic.

Bennett, Jr., W., G.M. Alliger, C.M. Colegrove, M.J. Garrity, and R. Beard. 2013 (In press). Mission Essential Competencies: A Novel Approach to Proficiency-Based Live, Virtual, and Constructive Readiness Training and Assessment. In Fundamental Issues in Defence Training and Simulation, edited by C. Best, G. Galanis, J. Kerry, and R. Sottilare. Burlington, Vt.: Ashgate.

Bloom, B.S. 1984. The 2 sigma problem: The search for methods of group instruction as effective as one-to-one tutoring. Educational Researcher 13(6): 4-16.

Clark, R.E., D. Feldon, J.J. Merrienboer, K. Yates, and S. Early. 2008. Cognitive Task Analysis. Pp. 577-594 in Handbook for Research for Educational Communications and Technology, edited by J.M. Spector, M.D. Merrill, J.J. G. van Merriënboer, and M.P. Driscoll. Mahwah, N.J.: Lawrence Erlbaum Associates.

Crandall, B., G. Klein, and R.R. Hoffman. 2006. Working Minds: A Practitioner's Guide to Cognitive Task Analysis. Cambridge, Mass.: MIT Press.

Fletcher, J.D. 2011. NS D-4260. DARPA Education Dominance Program: April 2010 and November 2010 Digital Tutor Assessments. Washington, D.C.: Institute for Defense Analyses.

Garrity, M. J., G.A. Alliger, R. Beard, W. Bennett, and C. Colegrove. 2012. Competency-Based Training within the Air Operations Center. A Blended Approach to Training—Job Analysis and Competency Models. Symposium at the American Psychological Association Annual Conference. Orlando, Fla.: American Psychological Society.

Hughes, MAJ S.E. 1994. The Evolution of the U.S. Army Infantry Squad: Where Do We Go From Here? Determining the Optimum Infantry Squad Organization for the Future. Fort Leavenworth, Kans.: School of Advance Military Studies, United States Army Command and General Staff College.

Katz, S., D. Allbritton, and J. Connelly. 2003. Going beyond the problem given: How human tutors use post-solution discussions to support transfer. International Journal of Artificial Intelligence in Education 13: 79-116.

Katz, S., G. Donnell, and H. Kay. 2000. An approach to analyzing the role and structure of reflective dialogue. International Journal of Artificial Intelligence in Education 11: 320-343.

Louie, M.A., and K.M. Carley. 2008. Balancing the criticisms: Validating multi-agent models of social systems. Simulation Modeling Practice and Theory 16(2): 242-256.

Marshall, Samuel L.A. 1950. The soldier's load and the mobility of a nation. Washington, D.C.: Combat Forces Press.

Meliza, L.L., S.L. Goldberg, and D.R. Lampton. 2007. RTO-TR-HFM-121-Part II AC/323(HFM-121)TP/62. After action review in simulation-based training. Section 8 in Part II of Final Report of Task Group TR-HFM-121. Available online http://ftp.rta.nato.int/public//PubFullText/RTO/TR/RTO-TR-HFM-121-PART-II///TR-HFM-121-Part-II-08.pdf. Accessed September 4, 2012.

Melody, MAJ P.E. 1990. AD-A225 438. The Infantry Rifle Squad: Size Is Not the Only Problem. Fort Leavenworth, Kans.: School of Advanced Military Studies, United States Army Command and General Staff College.

Miller, N.L., L.G. Shattuck, and P. Matsangas. 2011. Sleep and fatigue issues in continuous operations: A survey of US Army officers. Behavioral Sleep Medicine 9(1): 53 – 65.

NRC (National Research Council). 1997. Energy-Efficient Technologies for the Dismounted Soldier. Washington, D.C.: National Academy Press.

NRC. 2008. Behavioral Modeling and Simulation: From Individuals to Societies. Edited by G.L. Zacharias, J. MacMillan, and S.B. Van Hemel. Washington, D.C.: The National Academies Press.

NRC. 2009. Opportunities in Neuroscience for Future Army Applications. Washington, D.C.: The National Academies Press.

Rainey, MAJ J. E. 1998. Sharpening the Tip of the Spear: Is the Light Infantry Squad the Right Size for the Future Battlefield? Fort Leavenworth, Kans.: School of Advanced Military Studies, United States Army Command and General Staff College.

Schurig, I., S. Jarrett, W. Arthur Jr., R. Glaze, and M. Schurig. 2011. The Effectiveness of After–Action Reviews as a Training Method: A Meta–Analysis. Chicago, Ill.: 26th Annual Conference of the Society for Industrial and Organizational Psychology.

Thomas, M., H. Sing, G. Belenky, H. Holcomb, H. Mayberg, R. Dannals, H. Wagner, Jr., D. Thorne, K. Popp, L. Rowland, A. Welsh, S. Balwinski, and D. Redmond. 2000. Neural basis of alertness and cognitive performance impairments during sleepiness. I. Effects of 24 h of sleep deprivation on waking human regional brain activity. Journal of Sleep Research, 9(4): 335 – 352.

TRADOC (U.S. Army Training and Doctrine Command). 2010. TRADOC Pamphlet 525-3-5. The U.S. Army Functional Concept for Protection 2016-2028. Arlington, Va.: Headquarters, Department of the U.S. Army. Available online www.tradoc.army.mil/tpubs/pams/tp525-3-5.pdf. Accessed September 28, 2012.

U.S. Army. 1990. FM 21-18. Foot Marches. Arlington, Va.: Headquarters, Department of the Army. Available online http://www.everyspec.com/ARMY/FM-Field-Manual/FM_21-18_14353/. Accessed March 27, 2013.

U.S. Army. 2003. The Modern Warrior's Combat Load Dismounted Operations in Afghanistan, April - May 2003. Fort Leavenworth, Kans.: U.S. Army Center for Army Lessons Learned.

U.S. Army. 2010. Power and Energy Strategy White Paper. Fort Eustis, Va.: Army Capabilities Integration Center, U.S. Army Research, Development and Engineering Command.

Wesensten, N., G. Belenky, and T. Balkin. 2005a. Cognitive readiness in network-centric operations. Parameters: Journal of the US Army War College 35(1): 94 – 105.

Wesensten, N., W. Kilgore, and T. Balkin. 2005b. Performance and alertness effects of caffeine, dextroamphetamine, and modafinil during sleep deprivation. Journal of Sleep Research 14(3): 255-266.

Williamson, A., and A.-M. Feyer. 2000. Moderate sleep deprivation produces impairments in cognitive and motor performance equivalent to legally prescribed levels of alcohol intoxication. Occupational and Environmental Medicine 57(10): 649-655.

Wilson, M.A., W. Bennett Jr., S.G. Gibson, and G.M. Alliger, Editors. 2012. The Handbook of Work Analysis: The Methods, Systems, Applications and Science of Work Measurement in Organizations. Florence, Ky.: Routledge Academic.

Appendix A

Biographical Sketches of Committee Members

LTG Henry J. Hatch (Chair) has been an active volunteer with the National Research Council and several professional societies for the past 12 years. As an officer in the U.S. Army, he was the commanding general of the U.S. Army Corps of Engineers from 1988 to 1992. His missions included military construction and environmental engineering for the Army and Air Force and the Army's civil water resources program. After graduating from the U.S. Military Academy at West Point, he completed airborne and ranger training and earned a master's degree in geodetic science at the Ohio State University. General Hatch held combat engineer leadership positions in airborne and air assault units early in his Army career and commanded the 326th Engineer Battalion of the 101st Airborne Division in Vietnam. He was the Director of Combat Developments at the Army Engineer School. He oversaw military research, development, acquisition and civil construction as district engineer for the Corps of Engineers Nashville District and later as the division engineer for the Pacific Ocean Division before becoming the U.S. Army Chief of Engineers.

After retirement, General Hatch was the president and chief executive officer of Fluor Daniel Hanford, Inc., leading a $9 billion DOE environmental cleanup effort at the DOE Hanford site in Washington state. He also served as the chief operating officer of the American Society of Civil Engineers (ASCE) and was the chairman of the Law Companies Group, Inc., an international engineering and environmental services company. General Hatch is a registered professional engineer in the District of Columbia, a fellow of the Society of American Military Engineers, distinguished member of the American Society of Civil Engineers, and a member of the National Society of Professional Engineers, Tau Beta Pi, and Phi Kappa Phi.

W. Peter Cherry (NAE) is an independent consultant who recently retired as the chief analyst for the U.S. Army Future Combat Systems Program at Science Applications International Corporation (SAIC). He was responsible for analytic support to requirements analysis, performance assessment, and design trades.

Previously, Dr. Cherry was leader of the Integrated Simulation and Test Integrated Process Team, focusing on test and evaluation planning, the development of associated models and simulations, and the development of the Future Combat System of Systems Integration Laboratory. He has been a participant in the Future Combat Systems program since its inception, leading analysis and evaluation of concepts as a member of the Full Spectrum Team in the contract activities which preceded concept and technology development.

Before joining SAIC, he spent over 30 years with Vector Research Incorporated and its successor, the Altarum Institute. His professional career began in the field of maritime operations research in the Department of National Defence in Canada. He left that organization to obtain a Ph.D. in operations research at the University of Michigan, where he specialized in stochastic processes. Since the completion of his studies at the University of Michigan, he has focused on the development and application of operations research in the national security domain, primarily in the field of land combat. He contributed to the development and fielding of most of the major systems currently employed by the Army, from the Patriot missile system to the Apache helicopter, as well as the command control and intelligence systems currently in use such as ASAS and AFTAADS. In addition, he contributed to the creation of the Army's Manpower Personnel and Human Factors and Training Program (MANPRINT) and to the Army's Embedded Training Initiative. His recent research interests include peacekeeping operations and the development of transformational organizations and material.

Dr. Cherry received a B.A. in mathematics from the University of New Brunswick; an M.A. in mathematics from the University of Toronto; and an M.S. and a Ph.D. in industrial engineering from the University of Michigan.

Paul W. Glimcher is professor of neural sciences, economics and psychology at the New York University (NYU) Center for Neural Science and director of the university's Center for Neuroeconomics. He has achieved the following: A.B. – Princeton University, magna cum laude; Ph.D., University of Pennsylvania, Neuroscience; fellow of McKnight, Whitehall, Klingenstein and McDonnell Foundations. He is also investigator of the National Eye Institute, the National Institute of Mental Health, and the National Institute of Neurological Disorders and Stroke, founding president of the Society for Neuroeconomics, winner of the Margaret and Herman Sokol Faculty Award in the Sciences, 2003, and the winner of the NYU Distinguished (Lifetime Accomplishment) Teaching Award, 2006. He has been published in Nature, Science, Neuron Journal of Neurophysiology, American Economic Review, Games and Economic Behavior, Vision Research, Experimental Brain Research, and the MIT Encyclopedia of Cognitive Science. He is the author of Decisions, Uncertainty and, the Brain: The Science of Neuroeconomics and the winner of the American Association of Publishers Medical Sciences Book of the Year, 2003. Professor Glimcher's work has been covered by the Wall Street Journal, Time, Newsweek, The Los Angeles Times, Money Magazine, and New Scientist, and he has been heard on National Public Radio, the BBC and Fox News, among others.

Randal W. Hill, Jr. is the executive director for the University of Southern California (USC) Institute for Creative Technologies (ICT) and a research professor in computer science. While at ICT, he has been the director of applied research and transition, the deputy director of technology, and senior scientist. Previously, Dr. Hill was project leader and research scientist at USC's Information Sciences Institute and also held the positions of task manager, technical group leader and member of the technical staff at the California Institute of Technology Jet Propulsion Laboratory. He is a member of the Association for the Advancement of Artificial Intelligence (AAAI) and the American Society for Engineering Education (ASEE)

Dr. Hill served in the U.S. Army as a commissioned field artillery/military intelligence officer and was honorably discharged as captain in 1984. He received a B.S. from the U.S. Military Academy at West Point and an M.S. and Ph.D. in computer science from USC.

Robin L. Keesee is an independent consultant. A recently retired federal civil servant, he was vice director of the Joint IED Defeat Organization. As second-in-charge under senior general officers, he helped oversee the execution of the $3 billion to $4 billion per year mission. His emphasis was on the materiel initiatives, seeking technology and other countermeasures to IEDs drawing from across the Service and DOE labs, universities, defense contractors, and DARPA.

Earlier, Dr. Keesee had been the first deputy to the commanding general of the U.S. Army Research, Development and Engineering Command, Aberdeen Proving Ground (APG), Md., director of human research and engineering in the Army Research Laboratory, also at APG, and director of the Systems Research Laboratory of the Army Research Institute for the Behavioral and Social Sciences, Alexandria, Va. Dr. Keesee earned a B.S. in industrial engineering and a Ph.D. in human factors from Virginia Polytechnic Institute and State University.

Elliott D. Kieff (NAS/IOM) is the Albee Professor of Medicine and Microbiology and Molecular Genetics at the Channing Laboratory at Harvard University. Dr. Kieff has also held many distinguished academic positions at Harvard and the University of Chicago. He is the director of Infectious Diseases at the Brigham and Women's Hospital, the recipient of many honors from the American Academy of Arts and Sciences, and holds several patents, including a vaccine against the Epstein-Barr virus.

Dr. Kieff received a B.A. in chemistry from the University of Pennsylvania, a Ph.D. in microbiology from the University of Chicago, and an M.D. from Johns Hopkins University.

Jean MacMillan is the chief scientist of Aptima, Inc. She is a leading expert in understanding, maximizing, and assessing human performance in complex sociotechnical systems. Her 30-year career has spanned a broad range of accomplishments in simulation-based training, human-machine interaction, and user-centered system design. Dr. MacMillan's current research focuses on methods to increase the effectiveness of simulation-based training by linking training objectives to scenario design elements and performance measures. She recently led projects to develop reliable and valid performance measures for teams of F-16 pilots training in a distributed simulation facility and to design synthetic entities that function as team members for simulation-based training of teamwork skills.

Before joining Aptima in 1997, Dr. MacMillan was a senior scientist at BBN Technologies and a senior cognitive systems engineer at Alphatech (now BAE Systems). She is a frequent contributor and strategic advisor to workshops and expert panels on human engineering issues for organizations such as DARPA and the military services. Dr. MacMillan recently co-chaired a 3-year National Research Council study on military needs for social and organizational models, which resulted in the publication of *Behavioral Modeling and Simulation: From Individuals to Societies*.

Dr. MacMillan received a B.A. from Antioch College, an M.C.P. from Harvard University, and a Ph.D. in cognitive psychology from Harvard University. She is currently a member of the editorial board of the *Journal of Cognitive Engineering and Decision Making* and is associate editor for cognitive systems engineering for the on-line journal *Cognitive Technology*.

William L. Melvin is the director of the Sensors and Electromagnetic Applications Laboratory (SEAL) at the Georgia Tech Research Institute and an adjunct professor in Georgia Tech's Electrical and Computer Engineering Department. He has successfully developed and fostered major research thrusts within Georgia Tech centered on systems engineering, advanced signal processing, and high-fidelity modeling and simulation. His specific expertise includes digital signal processing with application to RF sensors, including adaptive signal processing for aerospace radar detection of airborne and ground moving targets, radar applications of detection and estimation theory, electronic protection, SIGINT, and synthetic aperture radar. He has authored over 150 articles in his areas of expertise and holds three patents on adaptive radar technology.

As director of SEAL, Dr. Melvin focuses a technology portfolio in excess of $36 million per year involving all aspects of sensor systems engineering, including: environmental characterization; antenna development; hardware and software design, implementation, test, and evaluation; advanced system concepts; signal processing; physics-based modeling and simulation; and field testing. Areas of recent special interest include deploying SAR-GMTI sensors on small UAVs; space-radar algorithm development and processing techniques; dismount detection and urban radar; multistatics; electronic protection; integrated air and missile defense; and, expeditionary force intelligence, surveillance, and reconnaissance.

Dr. Melvin is a fellow of the IEEE, with the following citation: "For contributions to adaptive signal processing methods in radar systems." He has served as a guest editor for several recent special sections appearing in the IEEE *Transactions on Aerospace and Electronic Systems* and acted as the technical co-chair of the 2001 IEEE Radar Conference and the 2004 IEEE Southeastern Symposium on System Theory. Dr. Melvin received a "Best Paper" award at the 1997 IEEE Radar Conference. He has provided tutorials and invited talks at a number of IEEE conferences and local IEEE section meetings, and he is a regular reviewer for several IEEE and IET journal publications. Dr. Melvin is the recent recipient of the 2006 IEEE AESS Young Engineer of the Year Award, the 2003 U.S. Air Force Research Laboratory Reservist of the Year Award, and the 2002 U.S. Air Force Materiel Command Engineering and Technical Management Reservist of the Year Award. Dr. Melvin received a Ph.D. in electrical engineering from Lehigh University in 1994, as well as MSEE and BSEE degrees (with high honors) from this same institution in 1992 and 1989, respectively.

Maj Gen Richard R. Paul, U.S. Air Force (ret.), is an independent consultant. He retired from the Boeing Company in 2007. Prior to Boeing, General Paul served 33 years in the U.S. Air Force, retiring in 2000.

During his 7-year Boeing career, General Paul served as a vice president in the Phantom Works, Boeing's centralized research and development organization that develops advanced technologies for Boeing's family of commercial aircraft and defense-related aerospace products and services. In 2006 and 2007, he concurrently served in the Office of the Chief Technology Officer (CTO), where his duties included executive management of Boeing's 2,000-person Technical Fellowship program and its External Technical Affiliations program, which served as the Boeing interface to dozens of professional societies.

During his 33-year Air Force career, General Paul served in three Air Force laboratories in New Mexico and Ohio; a product center in Massachusetts; two major command headquarters, in Nebraska and Ohio; Headquarters of the U.S. Air Force in the Pentagon; and a joint staff assignment in Nebraska. His assignments during the latter one-third of his career were aligned with the Air Force science and technology enterprise, where he served in his final assignment as the commander of the Air Force Research Laboratory headquartered in Dayton, Ohio.

Gen. Paul received a bachelor's degree in electrical engineering from the University of Missouri at Rolla (UMR) and a master's degree in electrical engineering from the Air Force Institute of Technology, and has been awarded a professional degree in electrical engineering by UMR. He is a distinguished graduate of the Air Command and Staff College at Maxwell Air Force Base in Alabama and the Naval War College at Newport, Rhode Island, and is a graduate of the Defense Systems Management College Program Management Course at Fort Belvoir, Virginia.

Richard Pew is a principal scientist at Raytheon BBN Technologies since 1976. At BBN Technologies Dr. Pew has been continuously involved in the analysis, design, and evaluation of systems in which human performance is a critical component. He has conducted studies of improved means of introducing human factors requirements in preliminary design. He has developed specific design recommendations for improved human interfaces in systems to be used by individuals with no knowledge of computers. He has also participated in experimental studies measuring human performance in computer-based systems. In addition, Dr. Pew has led and contributed to projects concerned with modeling and predicting human performance in applied settings. He has conducted studies concerned with understanding human performance and decision making and has continued his interests in human information processing.

Previously, Dr. Pew was a professor, associate professor, assistant professor and associate research psychologist and lecturer at the University of Michigan. His research at the university focused on basic and applied studies of human performance, including human information processing, perceptual motor performance, and the analysis and synthesis of manual control systems. In addition to his own work, he served as chairman or co-chairman of 15 Ph.D theses in these areas. From 1965 to 1998, he has served annually as course chairman for the University of Michigan Engineering Summer Conference on Human Factors Engineering. He continues to lecture in the course. Dr. Pew received a B.E.E. (electrical engineering) from Cornell University; an M.A. (psychology) from Harvard University; and a Ph.D. (psychology) from the University of Michigan.

M. Frank Rose is the chief technical officer for Radiance Technologies, Inc. Previously, he was the vice president for research. Prior to joining Radiance Technologies, he was director of the Science Directorate at the NASA Marshall Space Flight Center. Previous positions within the scientific community were as deputy director, Space Sciences Laboratory, NASA Marshall Space Flight Center; director, Space Power Institute, and professor, electrical engineering, Auburn University; and senior research scientist, Naval Surface Warfare Center. He has had a distinguished career involving progressively more responsible experience in performing and managing basic and applied research in the physical sciences and advanced technologies associated with space, shock wave physics, energy conversion, electronic warfare, directed energy technology, and space power technology. He has broad experience in planning, programming, coordinating, and implementing interdisciplinary R&D programs and he has international recognition in the field of advanced power technology and space environmental effects. He is a fellow of the IEEE, an associate fellow of the AIAA, and national associate of the National Academies. He was associate editor of the *Journal of Propulsion and Power* for 6 years and has been guest editor for several technical journals. He is the author/editor of five books, most dealing with high-power, high-speed phenomena; is the author of 160 technical papers in the open literature; and holds 12 patents, mostly in the area of advanced energy conversion. He is a past member of the NRC Board on Army Science and Technology and has participated in numerous BAST studies. He is a past member of the Scientific Advisory board for the Sandia National Laboratory. Dr. Rose holds a certificate in engineering in electrical engineering from the Clinch Valley College of the University of Virginia, a B.A. in physics from the University of Virginia, an M.Sc. in engineering physics, and a Ph.D. in engineering physics, from Pennsylvania State University.

Albert A. Sciarretta is president of CNS Technologies, Inc. (CNSTI), a company that consults on research and development, experimentation, chemical/biological defense, counterinsurgency operations, modeling and simulation, program development/management, and the assessment of the military utility of advanced technologies. His current personal efforts within CNSTI include serving as a senior research fellow at the Center for Technology and National Security Policy, National Defense University; supporting the program manager, Test and Evaluation/Science and Technology (T&E/S&T) in the Office of the Secretary of Defense (OSD); and serving as an on-call subject matter expert for an Independent Review Team for assessing technology programs identified by the assistant secretary of the Army for Acquisition, Logistics, and Technology (ASA[ALT]).

He has served as chief designer and director of an Office of the Secretary of Defense (OSD) experiment involving networked sensors in support of small-unit urban operations; an OSD demonstration of an integrated live-virtual-constructive simulation-based joint urban operations training environment; a U.S. Army experiment for micro-autonomous robots; and multiple Defense Advanced Research Projects Agency (DARPA) experiments of sensor and command and control (C2) technologies. His primary focus has been on small-unit operations in urban environments. His current efforts include assessing technologies for enhancing Department of Defense test ranges; developing a methodology and metrics for assessing the readiness of transitioning control of Afghanistan provinces to Afghanistan's central government; and designing and conducting an experiment for assessing the benefits of dynamic physical, information, and sociocognitive networks in small-unit dismounted operations.

Mr. Sciarretta is a retired Army officer. He has a B.S. degree in general engineering from the U.S. Military Academy and dual M.S. degrees in mechanical engineering and operations research from Stanford University. He previously served as a member of the National Research Council (NRC) Committee on Army Science and Technology for Homeland Defense: C4ISR; Army Unmanned Ground Vehicle Technology; Review of the Department of Defense Air and Space Systems Science and Technology Programs; and Advanced Energetic Materials and Manufacturing Technologies.

Ann E. Speed is a principal member of the technical staff at Sandia National Laboratories. Her background is in cognitive psychology areas of memory, analogy, training, language acquisition, and operant mechanisms of behavior. She has been at Sandia with the Cognitive Systems team for 10 years. Dr. Speed has over 20 years' experience conducting human subjects research and applying psychological principles to real-world problems. She has worked in areas as varied as combining synthetic perceptive systems with synthetic cognitive systems to enhance physical security, IED and terrorist network defeat, and computational models of group decisionmaking. Among other things, she is working on applying neurophysiological mechanisms of knowledge representation to computational modeling of human cognition in order to enable humanlike analogy making and learning in those computational models. Dr. Speed received a B.A. in psychology from the University of New Mexico, and an M.S. and Ph.D. in cognitive psychology from Louisiana State University.

LTG Joseph Yakovac is president of JVM LLC, and joined the Cohen Group as a senior counselor in July, 2008. General Yakovac retired from the U.S. Army in 2007, concluding more than 35 years of military service. His last assignment was director of the Army Acquisition Corps and military deputy to the Assistant Secretary of the Army for Acquisition, Logistics, and Technology. In those roles, he managed a dedicated team of military and civilian acquisition experts to make sure America's soldiers received state-of-the-art critical systems and support across a full spectrum of Army operations. He also provided critical military insight to the Department of Defense senior civilian leadership on acquisition management, technological infrastructure development, and systems management. General Yakovac also served as the program executive officer, Ground Combat Systems, and deputy for Systems Management and Horizontal Technology Integration.

After graduation from the U.S. Military Academy at West Point, he was commissioned in the infantry. He served as a platoon leader, executive officer, company and battalion commander in mechanized infantry units. General Yakovac earned an M.S. in mechanical engineering from the University of Colorado at Boulder before returning to West Point as an assistant professor. He is also a graduate of the Armor Officer Advanced Course, the Army Command and General Staff College, the Defense Systems Management College, and the Industrial College of the Armed Forces. He now teaches classes at the U.S. Military Academy, the Defense Management College, and the Naval Postgraduate School.

Appendix B

Committee Meetings

This appendix lists presentations given to the committee at its meetings and data-gathering sessions over the course of the study.

**FIRST COMMITTEE MEETING,JUNE 7-9 AND JUNE 13-14, 2011
WASHINGTON, DC, AND FORT BELVOIR, VA**

Study Statement of Task and Discussion
Dr. Scott Fish, Office of the Assistant Secretary of the Army for Acquisition, Logistics and Technology (ASA(ALT))

Tactical Small Units and Human Dimension Initiatives
Mr. Rickey Smith, ARCIC Forward, Army Capabilities Integration Center (ARCIC), U.S. Army Training and Doctrine Command (TRADOC)

PEO-Soldier Overview and Demonstrations
BG Nichols, Program Executive Officer – Soldier

RDECOM Science and Technology Program for Dismounted Soldiers
Dr. John P. Obusek, Natick Soldier Research, Natick Soldier Research, Development and Engineering Center

NVESD Overview and Demonstrations
Mr. David Randall, Systems Engineering, Night Vision Electronic Sensors Directorate (NVESD), U.S. Army Communications-Electronics Research, Development and Engineering Center (CERDEC)

Medical Research and Materiel Command, Research and Technology Programs
COL Carl Castro, U.S. Army Medical Research and Materiel Command

Maneuver Center of Excellence Overview
Mr. Donald M. Sando, Capabilities Development and Integration, U.S. Army Maneuver Center of Excellence (MCoE), Ft. Benning, GA

ARI Overview
Dr. Michelle Sams, U.S. Army Research Institute for the Behavioral and Social Sciences (ARI)

SECOND COMMITTEE MEETING, JULY 12-14, 2011
FORT BENNING, GA

MCoE Command Briefing
MAJ Ronald Sprang, U.S. Army MCoE

TRADOC Capabilities Manager (TCM)-Soldier
Mr. Patrick Berger, TRADOC Capabilities Manager-Soldier, TRADOC

Team Development Course—Leadership Reaction Course
CPT Lorang
1SG Divine

Directorate of Training
LTC Todd Zollinger, Directorate of Training, MCoE

Non-Commissioned Officers Academy
CSM Mark Horsley, Non-Commissioned Officers Academy, MCoE

Directorate of Training Development
Mr. Jay Brimstin, Directorate of Training Development, MCoE

Fire Team Training—Basic Rifle Marksmanship
CPT Konze, U.S. Army
1SG Cobb, U.S. Army

MCOE Army Research Laboratory Human Research and Engineering (ARL-HRED) Field Element
Dr. Elizabeth Redden, ARL-HRED

Army Research Institute for the Behavioral and Social Sciences
Ms. Jean Dyer, ARI

U.S. Army Infantry School
COL Jay Peterson, U.S. Army Infantry School

U.S. Army Armor School
COL Michael Wadsworth, U.S. Army Armor School

Maneuver Battle Lab Briefing
Mr. Edwin Davis, U.S. Army Maneuver Battle Lab

Outbrief to Capabilities Development and Integration Directorate
Mr. Don Sando, Capabilities Development and Integration Directorate, MCoE

SITE VISIT, SEPTEMBER 15-16, 2011
NATICK SOLDIER CENTER, NATICK, MA

Natick Soldier Research, Development, and Engineering Center (NSRDEC) Overview
Dr. John Obusek, NSRDEC

U.S. Army Research Institute of Environmental Medicine (USARIEM) Overview
LTC Robert Roussel, USARIEM

Overview of NSRDEC Human Dimension Research and Development Programs
Dr. Jason Augustyn, Human Dimension Science and Technology
Dr. Jeffrey Schiffman, Human Dimension Science and Technology
Ms. Betty Davis, Human Dimension Science and Technology
Dr. John Gassner, Material Science and Technology
Mr. Matthew Correa, Material Science and Technology
Mr. Michael Codega, Technology-Enabled Capability Demonstrations

Overview of USARIEM Human Dimension Research and Development Programs
Dr. Edward Zambraski, Soldier Physical and Cognitive Performance
Dr. Michael Sawka, Environmental Medicine
Dr. Reed Hoyt, Biomedical Modeling and Health Status Awareness
Dr. Andrew Young, Nutrition Science

Cognitive Performance in Operational and Environmental Contexts
NSRDEC and USARIEM Subject Matter Experts

Modeling and Analysis of Soldier and Small Unit Performance
NSRDEC and USARIEM Subject Matter Experts

Biomechanical and Physiological Aspects of Load Carriage and Mobility
NSRDEC and USARIEM Subject Matter Experts

Human Universal Load Carrier and NettWarrior
NSRDEC and USARIEM Subject Matter Experts

Nutrition and Soldier Performance
NSRDEC and USARIEM Subject Matter Experts

THIRD COMMITTEE MEETING, SEPTEMBER 20-22, 2011
ABERDEEN, MD

Communications-Electronics Research, Development, and Engineering Center (CERDEC) Overview
Daniel Buschmann, CERDEC

Army Research Laboratory Overview
Cary Chabalowski, ARL

ARL Weapons and Materials Research Directorate
Peter Plostins, Weapons and Materials Research Directorate, ARL

ARL Survivability/Lethality Analysis Directorate
Ms. Geri Kucinski, Survivability/Lethality Analysis Directorate, ARL

ARL Human Research and Engineering Directorate, Project Overview
Laurel Alender, HRED, ARL

Soldier Helmet Instrumentation
James Buxton, Aberdeen Test Center, CERDEC

Army Materiel Systems Analysis Activity Overview
Ginny Kistner. Army Materiel Systems Analysis Activity

SITE VISIT, OCTOBER 14, 2011
ARLINGTON, VA

ARI Overview
Dr. Michelle Sams, U.S. Army, Director, ARI

Enhancing Enlisted Personnel Management with Non-Cognitive Measures
Dr. Tonia Heffner, Personnel Assessment Research Unit, ARI

ARI Research in Cohesion and in Cross-Cultural Competence
Dr. Jay Goodwin, Foundational Science Research Unit, ARI

Training and Measurement for Small Tactical Units
Dr. Barbara Black, Training & Leader Development, ARI
Division

Appendix C

Army Terminology and Doctrine Relevant to Dismounted Soldier Missions

While this study was in progress, the Army made substantial changes in the preferred terminology for communicating doctrine and describing operations. With helpful advice from reviewers during the formal independent review of the draft report, the committee has aimed to employ the most current terminology and to use it in ways consistent with current Army doctrine. As an aid to readers, this appendix extracts key passages in order to define and explain terms that the committee views as particularly relevant to understanding the missions and tasks that the Army anticipates dismounted Soldiers, operating in small units, are likely to perform in future operations.

The following Army publications are the sources of the verbatim quotations given below:

- Army Doctrine Publication 1 (ADP 1), The Army, September 17, 2012. Available online http://armypubs.army.mil/doctrine/DR_pubs/dr_a/pdf/adp1.pdf (U.S. Army, 2012a).
- Army Doctrine Publication 3-0 (ADP 3-0), Unified Land Operations, October 10, 2011. Available online http://armypubs.army.mil/doctrine/DR_pubs/dr_a/pdf/adp3_0.pdf (U.S. Army, 2011).
- Army Doctrine Reference Publication 3-0 (ADRP 3-0), Unified Land Operations, May 16, 2012. Available online http://armypubs.army.mil/doctrine/DR_pubs/dr_a/pdf/adrp3_0.pdf. (U.S. Army, 2012b).

ADP 3-0 is a concise, high-level introduction, in 19 pages, to the Army's operational concept of "unified land operations," which supersedes the previous operational concept of "full spectrum operations." ADRP 3-0, which runs to more than 60 pages of introduction, chapters, and glossary, "expands the discussion of the foundations and tenets of unified land operations, as well as the operational framework found in ADP 3-0." (U.S. Army, 2012b, p. v).

THE ARMY PROVIDES LANDPOWER TO WIN IN THE LAND DOMAIN

From U.S. Army, 2012a, Page 1-1

U.S. forces operate in the air, land, maritime, space, and cyberspace domains.... War begins and ends based upon how it affects the land domain. Air, maritime, space, and cybernetic power affect the land domain indirectly; landpower is usually the arbiter of victory. The Army provides the United States with the landpower to prevent, shape, and win in the land domain.

THE LAND DOMAIN

1-1. The distinguishing characteristic of the land domain is the presence of humans in large numbers.... Humans live on the land and affect almost every aspect of land operations. Soldiers operate among populations, not adjacent to them or above them. They accomplish missions face-to-face with people, in the midst of environmental, societal, religious, and political tumult. Winning battles and engagements is important but alone is usually insufficient to produce lasting change in the conditions that spawned conflict. Our effectiveness depends on our ability to manage populations and civilian authorities as much as it does on technical competence employing equipment. Managing populations before, during, and after all phases of the campaign normally determines its success or failure. Soldiers often cooperate, shape, influence, assist, and coerce according to the situation, varying their actions to make permanent the otherwise temporary gains achieved through combat.

THE RANGE OF MILITARY OPERATIONS

From U.S. Army, 2012b, Page 1-6

1-38. Military operations vary in purpose, scale, risk, and intensity (see JP 3-0). They include relatively benign, routine, and recurring military operations in peacetime; specific combat and noncombat responses to contingencies and crises as they occur; and less frequent, large-scale combat operations typical of wartime conditions. Army forces are designed, organized, equipped, and trained to accomplish many military operations. Table 1-1 lists examples of military operations. (See JP 1 for a discussion of the range of military operations.)

Table 1-1. Examples of operations and their applicable doctrine

Arms control and disarmament (JP 3-0)	Large-scale combat (FM 3-90)
Civil support (JP 3-28 and FM 3-28)	Noncombatant evacuation (JP 3-68)
Civil-military operations (JP 3-57)	Peace operations (JP 3-07.3)
Combating terrorism (JP 3-07.2)	Raid (FM 3-90)
Combating weapons of mass destruction (JP 3-40)	Recovery operations (JP 3-50 and FM 3-50.1)
Counterinsurgency (JP 3-24 and FM 3-24)	Security force assistance (AR 12-1 and FM 3-07.1)
Enforcement of sanctions (JP 3-0)	Show of force (JP 3-0)
Foreign humanitarian assistance (JP 3-29)	Stability tasks (FM 3-07)
Foreign internal defense (JP 3-22 and FM 3-05.2)	Strike (JP 3-0)
Homeland defense (JP 3-27 and FM 3-28)	Unconventional warfare (JP 3-05 and FM 3-05)

["JP" refers to a document in the Joint Publication series; "FM" and "AR" refer to Army documents.]

APPENDIX C

OVERSEAS ARMY OPERATIONS COMBINE OFFENSIVE, DEFENSIVE, AND STABILITY TASKS IN DECISIVE ACTIONS

From U.S. Army, 2012a, Pp. 1-2 to 1-3

LAND OPERATIONS

1-4. Land combat against an armed adversary is an intense, lethal human activity. Its conditions include complexity, chaos, fear, violence, fatigue, and uncertainty. The battlefield often teems with noncombatants and is crowded with infrastructure. In any conflict, Soldiers potentially face regular, irregular, or paramilitary enemy forces that possess advanced weapons and rapidly communicate using cellular devices. Our enemies will employ terror, criminal activity, and every means of messaging to further complicate our tasks. To an ever-increasing degree, activities in cyberspace and the information environment are inseparable from ground operations. Successful land combat requires protected friendly networks (wired and wireless) while exploiting or degrading the enemy's networks. The information environment, our use of it, and inform and influence activities continues to increase. Because the land environment is so complex, the potential for unintended consequences remains quite high. In the end, it is not the quality of weapons, but the quality of Soldiers employing them that determines mission success.

1-5. Any mission can rapidly become a combination of combat, governance, and civil security. Most of our missions require combinations of lethal and nonlethal actions. This is inherent in the nature of land operations, usually conducted in the midst of noncombatants. When called upon, Soldiers accomplish nonlethal missions such as disaster relief and humanitarian assistance quickly and effectively. Regardless, our combat capability often underwrites our ability to provide assistance. Nobody in or outside the military profession should mistake the Army for anything other than a force organized, equipped, and trained for winning the Nation's wars.

1-6. *Unified Land Operations* is the title of the Army's basic operational doctrine, ADP 3-0. It emphasizes the necessity of synchronizing our capabilities with the other Services (joint), other government agencies (interagency), other international government partners (intergovernmental), and military forces from partner nations (multinational). The basic premise of unified land operations is that Army forces combine offensive tasks, defensive tasks, stability tasks, and defense support of civil authorities (DSCA) in concert with joint, interagency, intergovernmental, and multinational partners. Army operations conducted overseas combine offensive, defensive, and stability tasks. Within the United States, we support civil authorities through DSCA. If hostile powers threaten the homeland, we combine defensive and offensive tasks with DSCA. The effort accorded to each task is proportional to the mission and varies with the situation. We label these combinations *decisive action* because of their necessity in any campaign.

From U.S. Army, 2012b, Pages 2-2, 2-4

DECISIVE ACTION

2-9. Army forces demonstrate the Army's core competencies through *decisive action*—**the continuous, simultaneous combinations of offensive, defensive, and stability or defense support of civil authorities tasks**. In unified land operations, commanders seek to seize, retain, and exploit the initiative while synchronizing their actions to achieve the best effects possible. Operations conducted outside the United States and its territories simultaneously combine three elements—offense, defense, and stability. Within the United States and its territories, decisive action combines the elements of defense support of civil authorities and, as required, offense and defense to support homeland defense. . . .

2-18. Decisive action requires simultaneous combinations of offense, defense, and stability or defense support of civil authorities tasks. Table 2-1 lists the tasks associated with each element and the purposes of each task. Each task has numerous associated subordinate tasks. When combined with who (unit), when (time), where (location), and why (purpose), the tasks may become mission statements.

Table 2-1. Tasks of decisive action

Offense	***Defense***
Tasks:	Tasks:
• Movement to contact • Attack • Exploitation • Pursuit	• Mobile defense • Area defense • Retrograde
Purposes:	Purposes:
• Dislocate, isolate, disrupt, and destroy enemy forces • Seize key terrain • Deprive the enemy of resources • Develop intelligence • Deceive and divert the enemy • Create a secure environment for stability tasks	• Deter or defeat enemy offense • Gain time • Achieve economy of force • Retain key terrain • Protect the populace, critical assets, and infrastructure • Develop intelligence
Stability	***Defense Support of Civil Authorities***
Tasks:	Tasks:
• Establish civil security (including security force assistance) • Establish civil control • Restore essential services • Support to governance • Support to economic and infrastructure development	• Provide support for domestic disasters • Provide support for domestic chemical, biological, radiological, and nuclear incidents • Provide support for domestic civilian law enforcement agencies • Provide other designated support
Purposes:	Purposes:
• Provide a secure environment • Secure land areas • Meet the critical needs of the populace • Gain support for host-nation government • Shape the environment for interagency and host-nation success	• Save lives • Restore essential services • Maintain or restore law and order • Protect infrastructure and property • Maintain or restore local government • Shape the environment for interagency success

Offensive Tasks

2-19. An offensive task is a task conducted to defeat and destroy enemy forces and seize terrain, resources, and population centers. Offensive tasks impose the commander's will on the enemy. In combined arms maneuver, the offense is a task of decisive action. Against a capable, adaptive enemy, the offense is the most direct and a sure means of seizing, retaining, and exploiting the initiative to gain physical and psychological advantages and achieve definitive results. In the offense, the decisive operation is a sudden, shattering action against an enemy weakness that capitalizes on speed, surprise, and shock. If that operation does not destroy the enemy, operations continue until enemy forces disintegrate or retreat to where they no longer pose a threat. Executing offensive tasks compels the enemy to react, creating or revealing additional weaknesses that the attacking force can exploit. (See Army tactics doctrine for a detailed discussion on offensive tasks.)

Defensive Tasks

2-20. A defensive task is a task conducted to defeat an enemy attack, gain time, economize forces, and develop conditions favorable for offensive or stability tasks. Normally the defense alone cannot achieve a decision. However, it can set conditions for a counteroffensive or counterattack that enables Army forces to regain the initiative. Defensive tasks can also establish a shield behind which wide area security can progress. Defensive tasks are a counter to the enemy offense. They defeat attacks, destroying as much of the attacking enemy as possible. They also preserve and maintain control over land, resources, and populations. The purpose of defensive tasks is to retain terrain, guard populations, and protect critical capabilities against enemy attacks. Commanders can conduct defensive tasks to gain time and economize forces so offensive tasks can be executed elsewhere. (See Army tactics doctrine for a detailed discussion on defensive tasks.)

Stability Tasks

2-21. Stability is an overarching term encompassing various military missions, tasks, and activities conducted outside the United States in coordination with other instruments of national power to maintain or reestablish a safe and secure environment, provide essential governmental services, emergency infrastructure reconstruction, and humanitarian relief. (See JP 3-0.) Army forces conduct stability tasks during both combined arms maneuver and wide area security. These tasks support a host-nation or an interim government or part of a transitional military authority when no government exists. Stability tasks involve both coercive and constructive actions. They help to establish or maintain a safe and secure environment and facilitate reconciliation among local or regional adversaries. Stability tasks can also help establish political, legal, social, and economic institutions while supporting the transition to legitimate host-nation governance. Stability tasks cannot succeed if they only react to enemy initiatives. Stability tasks must maintain the initiative by pursuing objectives that resolve the causes of instability. (See Army doctrine on stability tasks.)

CHARACTERIZING THE OPERATIONAL ENVIRONMENT WITH OPERATIONAL AND MISSION VARIABLES

From U.S. Army, 2012b, Page 2

THE OPERATIONAL ENVIRONMENT

7. The operational environment is a composite of the conditions, circumstances, and influences that affect the employment of capabilities and bear on the decisions of the commander (JP 1-02[1]). Army leaders plan, prepare, execute, and assess operations by analyzing the operational environment in terms of the operational variables and mission variables. The operational variables consist of political, military, economic, social, information, infrastructure, physical environment, time (known as PMESII-PT). The mission variables consist of mission, enemy, terrain and weather, troops and support available, time available, civil considerations (known as METT-TC). How these variables interact in a specific situation, domain (land, maritime, air, space, or cyberspace), area of operations, or area of interest describes a commander's operational environment but does not limit it. No two operational environments are identical, even within the same theater of operations, and every operational environment changes over time. Because of this, Army leaders consider how evolving relevant operational or mission variables affect force employment concepts and tactical actions that contribute to the strategic purpose.

THE ARMY'S CORE COMPETENICES: COMBINED ARMS MANEUVER AND WIDE AREA SECURITY

From U.S. Army, 2012b, Pages 2-8 - 2-10

ARMY CORE COMPETENCIES

2-31. Army forces demonstrate their core competencies of combined arms maneuver and wide area security by combining offensive, defensive, and stability or defense support of civil authorities tasks simultaneously. As part of a combined arms force within unified land operations, Army forces accept prudent risk to create opportunities to achieve decisive results. They employ synchronized action of lethal and nonlethal effects, proportional to the mission and informed by an understanding of an operational environment. Mission command that conveys commander's intent guides the adaptive use of Army forces.

2-32. Although distinct by definition, combined arms maneuver and wide area security are inseparable and simultaneous. Combined arms maneuver and wide area security provide the Army a focus for decisive action as well as a construct for understanding how Army forces use combined arms to achieve success in this contest of wills. As core competencies, combined arms maneuver and wide area security uniquely define what the Army provides to

[1]"JP 1-02" refers to Joint Publication 1-02, Department of Defense Dictionary of Military and Associated Terms, Directorate for Joint Force Development (J-7), Joint Staff (DoD, 1994). In ADRP 3-0, operational environment is defined and discussed in paragraphs 1-2 through 1-16 (U.S. Army, 2012b). That discussion includes the PMESII-PT operational variables (paragraph 1-9) and the METT-TC mission variables (paragraph 1-10).

the joint force commander. Additionally, the Army is organized and equipped to support the joint force commander through combined arms to cover vast distances for extended periods. The Army works to integrate all available instruments to unified action partners to achieve the desired outcome.

2-33. Combined arms maneuver and wide area security are not tasks. They provide an operational context to assist a commander and staff in determining an operational approach and to combine tasks of decisive action into a coherent operation that assigns missions to subordinates. Forces execute these missions to defeat or destroy enemy forces, and seize or control areas vital to accomplishing their missions, while protecting civilians, infrastructure, and themselves. While all operations consist of simultaneous combined arms maneuver and wide area security in various proportions, most tactical tasks will be predominantly characterized by one or the other. The preponderant core competency determines the choice of defeat or stability mechanisms to describe how friendly forces accomplish the assigned mission. Generally, defeat mechanisms are appropriate for combined arms maneuver, while stability mechanisms are best suited for wide area security.

Combined Arms Maneuver

2-34. *Combined arms maneuver* is the application of the elements of combat power in unified action to defeat enemy ground forces; to seize, occupy, and defend land areas; and to achieve physical, temporal, and psychological advantages over the enemy to seize and exploit the initiative (ADP 3-0). Physical advantages may include control of key terrain, population centers, or critical resources and enablers. Temporal advantages enable Army forces to set the tempo and momentum of operations and decide when to fight so the enemy loses the ability to respond effectively. Psychological advantages impose fear, uncertainty, and doubt on the enemy, which serves to dissuade or disrupt the enemy's further planning and action.

2-35. Combined arms maneuver exposes enemies to friendly combat power from unexpected directions and denies them the ability to respond effectively. Combined arms maneuver throws the enemy off balance, follows up rapidly to prevent recovery, and destroys the enemy's will to fight. In addition, forces conducting combined arms maneuver threaten enemies indirectly, causing them to reveal their intentions and expose hidden vulnerabilities. Combined arms maneuver primarily employs defeat mechanisms against enemies and is dominated by offensive and defensive tasks.

2-36. **A *defeat mechanism* is a method through which friendly forces accomplish their mission against enemy opposition**. Army forces at all echelons use combinations of four defeat mechanisms: destroy, dislocate, disintegrate, and isolate. Applying focused combinations produces complementary and reinforcing effects not attainable with a single mechanism. Used individually, a defeat mechanism achieves results proportional to the effort expended. Used in combination, the effects are likely to be both synergistic and lasting. When commanders destroy, they apply lethal combat power on an enemy capability so that it can no longer perform any function. The enemy cannot restore it to a usable condition without being entirely rebuilt. Commanders dislocate by employing forces to obtain significant positional advantage, rendering the enemy's dispositions less valuable, perhaps even irrelevant. Disintegrate means to disrupt the enemy's command and control system, degrading its ability to conduct operations. This action leads to a rapid collapse of the enemy's capabilities or will to fight. When commanders isolate, they deny an enemy or adversary access to capabilities that enable the exercise of coercion, influence, potential advantage, and freedom of action . . . .

Wide Area Security

2-39. *Wide area security* is the application of the elements of combat power in unified action to protect populations, forces, infrastructure, and activities; to deny the enemy positions of advantage; and to consolidate gains in order to retain the initiative (ADP 3-0). Army forces conduct security tasks to provide the joint force commander with reaction time and maneuver space. Additionally, these forces defeat or fix the enemy before the enemy can attack, thus allowing the commander to retain the initiative.

2-40. As part of unified land operations, Army forces may assist the development of host-nation security forces, a viable market economy, the rule of law, and an effective government by establishing and maintaining security in an area of operations. The goal is a stable civil situation sustainable by host-nation assets without Army forces. Security, the health of the local economy, and the capability of self-government are related. Without security, the local economy falters, populations feel unsecure, and enemy forces gain an advantage. A functioning economy provides employment and reduces the dependence of the population on the military for necessities. Security and economic stability precede an effective and stable government.

2-41. Wide area security includes the minimum essential stability tasks as part of decisive action. Army forces perform five primary stability tasks:

- Establish civil security, including security force assistance.
- Establish civil control.
- Restore essential services.
- Support governance.
- Support economic and infrastructure development.

2-42. The combination of stability tasks conducted during operations depends on the situation. In some operations, the host nation can meet most or all of the population's requirements. In those cases, Army forces work with and through host-nation authorities. Commanders use civil affairs operations to mitigate how the military presence affects the populace and vice versa. Conversely, Army forces operating in a failed state may need to support the well-being of the local populace. That situation requires Army forces to work with civilian organizations to restore basic capabilities. Again, civil affairs operations prove essential in establishing trust between Army forces and civilian organizations required for effective, working relationships.

2-43. **A *stability mechanism* is the primary method through which friendly forces affect civilians in order to attain conditions that support establishing a lasting, stable peace.** As with defeat mechanisms, combinations of stability mechanisms produce complementary and reinforcing effects that accomplish the mission more effectively and efficiently than single mechanisms do alone.

2-44. The four stability mechanisms are compel, control, influence, and support. Compel means to use, or threaten to use, lethal force to establish control and dominance, effect behavioral change, or enforce compliance with mandates, agreements, or civil authority. Control involves imposing civil order. Influence means to alter the opinions, attitudes, and ultimately behavior of foreign friendly, neutral, adversary, and enemy populations through inform and influence activities, presence, and conduct. Support is to establish, reinforce, or set the conditions necessary for the instruments of national power to function effectively.

THE ELEMENTS OF COMBAT POWER AND THE SIX WARFIGHTING FUNCTIONS

From U.S. Army, 2012b, Pages 3-1 - 3-6

3-1. Combined arms maneuver and wide area security, executed through simultaneous offensive, defensive, stability, or defense support of civil authorities tasks, require continuously generating and applying combat power, often for extended periods. ***Combat power*** **is the total means of destructive, constructive, and information capabilities that a military unit or formation can apply at a given time**. Army forces generate combat power by converting potential into effective action.

3-2. To execute combined arms operations, commanders conceptualize capabilities in terms of combat power. Combat power has eight elements: leadership, information, mission command, movement and maneuver, intelligence, fires, sustainment, and protection. The Army collectively describes the last six elements as the warfighting functions. Commanders apply combat power through the warfighting functions using leadership and information. (See figure 3-1.)

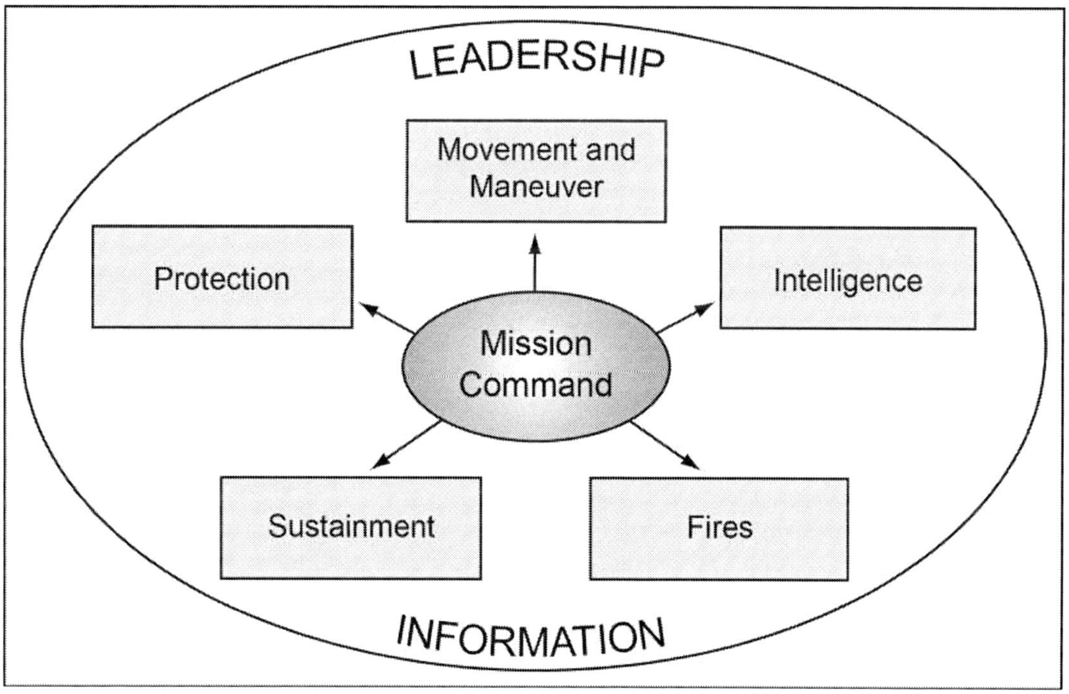

FIGURE C-1 The elements of combat power. SOURCE: U.S. Army, 2012b.

3-6. Commanders use the warfighting functions to help them exercise command and to help them and their staffs exercise control. A warfighting function is a group of tasks and systems (people, organizations, information, and processes) united by a common purpose that commanders use to accomplish missions and training objectives. All warfighting functions possess scalable capabilities to mass lethal and nonlethal effects. The Army's warfighting functions link directly to the joint functions.

3-7. **The *mission command warfighting function* is the related tasks and systems that develop and integrate those activities enabling a commander to balance the art of command and the science of control in order to integrate the other warfighting functions.** Commanders, assisted by their staffs, integrate numerous processes and activities within the headquarters and across the force as they exercise mission command. . . .

3-15. **The *movement and maneuver warfighting function* is the related tasks and systems that move and employ forces to achieve a position of relative advantage over the enemy and other threats.** Direct fire and close combat are inherent in maneuver. The movement and maneuver warfighting function includes tasks associated with force projection related to gaining a position of advantage over the enemy. Movement is necessary to disperse and displace the force as a whole or in part when maneuvering. Maneuver is the employment of forces in the operational area. It works through movement and with fires to achieve a position of advantage relative to the enemy to accomplish the mission. Commanders use maneuver for massing the effects of combat power to achieve surprise, shock, and momentum. Effective maneuver requires close coordination with fires. Both tactical and operational maneuver require sustainment support. The movement and maneuver warfighting function includes the following tasks:

- Deploy.
- Move.
- Maneuver.
- Employ direct fires.
- Occupy an area.
- Conduct mobility and countermobility operations.
- Conduct reconnaissance and surveillance.
- Employ battlefield obscuration. . . .

3-17. **The *intelligence warfighting function* is the related tasks and systems that facilitate understanding the enemy, terrain, and civil considerations.** This warfighting function includes understanding threats, adversaries, and weather. It synchronizes information collection with the primary tactical tasks of reconnaissance, surveillance, security, and intelligence operations. Intelligence is driven by commanders and is more than just collection. Developing intelligence is a continuous process that involves analyzing information from all sources and conducting operations to develop the situation. The warfighting function includes specific intelligence and communication structures at each echelon. The intelligence warfighting function includes the following tasks:

- Support force generation.
- Support situational understanding.
- Provide intelligence support to targeting and information capabilities.
- Collect information.

3-18. The intelligence warfighting function provides specific intelligence capabilities and communication structures at each echelon from the national level through the tactical level. These capabilities and structures include intelligence organizations, systems, and procedures for generating intelligence reports. They also include products, visualization aides, situational understanding and awareness products, and other critical information products. Effective communication connectivity and automation are essential components of this architecture. (FM 2-0 discusses the intelligence warfighting function.). . . .

3-19. **The *fires warfighting function* is the related tasks and systems that provide collective and coordinated use of Army indirect fires, air and missile defense, and joint fires through the targeting process**. Army fires systems deliver fires in support of offensive and defensive tasks to create specific lethal and nonlethal effects on a target. The fires warfighting function includes the following tasks:

- Deliver fires.
- Integrate all forms of Army, joint, and multinational fires.
- Conduct targeting. . . .

3-20. **The *sustainment warfighting function* is the related tasks and systems that provide support and services to ensure freedom of action, extend operational reach, and prolong endurance**. The endurance of Army forces is primarily a function of their sustainment. Sustainment determines the depth and duration of Army operations. It is essential to retaining and exploiting the initiative. Sustainment provides the support necessary to maintain operations until mission accomplishment. The sustainment warfighting function includes the following tasks:

- Conduct logistics.
- Provide personnel services.
- Provide health service support. . . .[2]

3-26. **The *protection warfighting function* is the related tasks and systems that preserve the force so the commander can apply maximum combat power to accomplish the mission**. Preserving the force includes protecting personnel (combatants and noncombatants) and physical assets of the United States and multinational military and civilian partners, to include the host nation. The protection warfighting function enables the commander to maintain the force's integrity and combat power. Protection determines the degree to which potential threats can disrupt operations and then counters or mitigates those threats. Protection is a continuing activity; it integrates all protection capabilities to safeguard bases, secure routes, and protect forces. To ensure maintenance of the critical asset list and the defended asset list and associated resourcing of fixed sites and forces against air and indirect fire threats, air and missile defense participates in meetings geared to protection activities. The protection warfighting function includes the following tasks:

- Conduct operational area security.
- Employ safety techniques (including fratricide avoidance).
- Implement operations security.
- Implement physical security procedures.
- Provide intelligence support to protection.
- Implement information protection.
- Apply antiterrorism measures.
- Conduct law and order.
- Conduct survivability operations.

---

[2]Paragraphs 3-21 through 3-25 of ADRP 3-0 describe the sustainment functions under logistics, personnel services, and health service support, including references to more detailed Army documents on each of these sustainment tasks (U.S. Army, 2012b).

- Provide force health protection.
- Conduct chemical, biological, radiological, and nuclear operations.
- Provide explosive ordnance disposal and protection support.
- Coordinate air and missile defense.
- Conduct personnel recovery operations.
- Conduct internment and resettlement.

REFERENCES

DoD (Department of Defense). 1994. Joint Pub 1-02. Department of Defense Dictionary of Military and Associated Terms. Washington, D.C.: Department of Defense.

U.S. Army. 2011. ADP 3-0. Unified Land Operations. Available online http://armypubs.army.mil/doctrine/DR_pubs/dr_a/pdf/adp3_0.pdf. Accessed February 6, 2013.

U.S. Army. 2012a. ADP 1. The Army (Incl C1). Available online http://armypubs.army.mil/doctrine/DR_pubs/dr_a/pdf/adp1.pdf. Accessed February 6, 2013.

U.S. Army. 2012b. ADRP 3-0. Unified Land Operations. Available online http://armypubs.army.mil/doctrine/DR_pubs/dr_a/pdf/adrp3_0.pdf. Accessed February 6, 2013.

Appendix D

History and Status of Design for the Soldier as a System

This appendix summarizes efforts taken by the Army over the last two decades to analyze Soldier requirements using a systems perspective, including the 1991 study by the Army Science Board (ASB) and subsequent Soldier-as-a-system program developments (ASB, 1991).

THE 1991 ARMY SCIENCE BOARD STUDY

In December 1991, the ASB published a report entitled *Soldier as a System*, which stressed the importance of treating the Soldier in a systems context (ASB, 1991). Since then, development efforts for the Soldier have often been referred to as "Soldier systems" or as supporting Soldier-as-a-system. Given the complexity of the systems being considered for the Soldier of the future, and depending on whether or not one identifies Soldier devices (e.g., weapons, body armor, night vision goggles, sensors) as components, subsystems, or systems; the Soldier may also be considered as a system of systems—a collection of task-oriented or dedicated systems that integrate their capabilities to create a new, more complex system that offers more functionality and performance than simply the sum of the constituent systems. In a similar manner, the tactical small unit (TSU) can be viewed as not merely a formation but as an organization or, better yet, a system-of-systems, which should be optimized to efficiently and effectively accomplish core and supporting missions in a constrained environment.

The 1991 ASB report, while 20 years old, still has a number of findings and recommendations that the committee believes remain applicable generally to the Army's future unified land operations and specifically to the subject of this study: ensuring that future dismounted TSUs and Soldiers have decisive overmatch across the gamut of those operations. Even though the ASB report dealt primarily with the multiple facets of materiel-related capabilities and the need for an integrated perspective, today there is broad recognition that multiple facets of the human dimension, in addition to the materiel dimension, are critical to this broad range of missions and operating environments. Thus, the need to treat the Soldier as a system—a system with both materiel and human dimensions—is even more critical today than it was at the time the ASB report was written. In particular, the following excerpt from the Executive Summary of that report seems appropriate to the contemporary environment:

> All the multiple components of the Soldier System—the programs, organization, systems, technologies, and soldier types—interact and interrelate. The justification for treating the Soldier System as a major system with integrated management perspective, although potent, must not overlook the difficulties of such an approach. The Soldier System Manager must manage complexity of a high order. (ASB, 1991, p. 1)

Two themes in the 1991 ASB report seem particularly relevant to today's environment: the need for and importance of (1) an integrated architecture design and (2) a systems engineering methodology. The ASB report defined architecture as follows:

> . . . a substantive definition of the elements within the Soldier System and a definition of how each of these elements is to interface with each other; a substantive definition of the primary elements outside the Soldier System with which the soldier must deal and a companion definition of these required interfaces; and a reasonably complete definition of the expected implementation concepts for fielding, both in timing of individual element introduction and in the ability/inability to use in part or mix/matched with existing inventory items. (U.S. Army, 1991, p. 33)

The report defined system engineering as follows:

> . . . System engineering establishes the desired requirements; defines a system architecture specifying form, fit, and function of the elements to ensure compatibility and interchangeability of the parts; and maintains the configuration in documentation available to all contributors to the development and provisioning activities. (ASB, 1991, p. 34)

The report went on to observe that both a design architecture and a systems engineering methodology were essential to realizing the system Soldier and went on to make a number of recommendations for pursuing these critical elements.

FOLLOW-ON TO SOLDIER AS A SYSTEM

The recommendations of the ASB report are supported by a subsequent review, *Objective Force Warrior Technology Assessment*, chartered in 2000 by the Deputy Assistant Secretary of the Army for Research and Technology.[1] The charter to the Independent Review Team (IRT) that conducted the study described the Objective Force Warrior as possessing the agility and versatility to operate with overmatch across the spectrum of conflict, environmental complexity, and mission set: offense, defense, stability, and support. It is interesting to note the similarity of this charter to the Statement of Task given to the current committee.

The IRT made recommendations related to power, weight, lethality, human performance, training, and integration. In particular, the IRT concluded as follows:

- Early integration avoids suboptimal science and technology (S&T) investment,
- System-level design is needed to determine early S&T investment, and
- An organization with the Objective Force Warrior systems design capability could not be identified among the presenters.

The IRT also assessed systems integration and modeling to be in need of redirection and model integration as needing additional funding. The IRT's recommendations were as follows:

---

[1]Personal communications between Ed Brady, chair of the IRT for Dr. Andrews, and Peter Cherry, committee member, who was also a member of the team.

- Implement an integrated S&T approach, to include:
  —Creation of a warrior systems design office
  —Provision of adequate funding, and
  —Development of a virtual prototype of the warrior system.
- Energize the Soldier Integrated Concept Team and strengthen S&T input.

INDICATORS OF FAILURE TO INTEGRATE TSU AND SOLDIER DECISIVE OVERMATCH CAPABILITIES

It was disappointing—at least to the current committee—to learn that the Army's responses to the ASB recommendations for the Soldier in 1991 and similar recommendations from the IRT in 2000 have not been successfully integrated in the way that dismounted TSUs and Soldiers are prepared for the missions they face. If anything, the current and projected demands upon the dismounted Soldier and the TSU are greater and more critical tactically, operationally, and strategically. The importance of implementing a systems approach and creating a single management authority to equip and prepare the dismounted TSU and the Soldier cannot be overstated. Nevertheless, despite numerous mentions of the "Soldier as a system" as being key to Soldier and TSU performance and consequent warfighting effectiveness since at least the 1991 ASB report, the Army has not adequately applied systems engineering discipline to either the Soldier or the dismounted TSU.

Although the Army's combat development community (e.g., the U.S. Army Training and Doctrine Command) has identified many physical and cognitive performance capabilities that would enhance Soldier and TSU enhanced warfighting effectiveness, even a cursory comparison of desired to currently fielded force capabilities identifies numerous capability gaps. Given the range of TSU and Soldier capability gaps to be addressed and the complex solution space of potential Doctrine, Organization, Training, Materiel, Leadership and Education, Personnel and Facilities (DOTMLPF) solutions, the Army should be applying systems engineering discipline to close these gaps, just as it does for its major platform systems and other systems-of-systems that currently have decisive overmatch. However, DOTMLPF enhancements for individual Soldiers and TSUs appear to be based on independent efforts ("eaches") rather than on integrated systems engineering. This issue is not limited to Army combat developers; the materiel development community —comprising the Army Research, Development and Engineering Command and the Program Executive Offices and program managers under the Army Acquisition Executive—also exhibit this limitation.

The committee believes that the following problems and failures in recent operations—Operation Enduring Freedom (OEF) and Operation Iraqi Freedom (OIF)—exemplify the lack of success in applying adequate systems engineering discipline.

Network Integration

In dismounted operations, Soldiers and TSUs are often not integrated into the Army network. One result is that they are too often surprised in tactical situations, resulting in unnecessary casualties. Dismounted TSUs and Soldiers lack sufficient timely situational understanding of the locations of their supporting assets, the enemy, and noncombatants.

Battery Proliferation

Numerous batteries of varying sizes, shapes, and power outputs must be used by dismounted Soldiers and TSUs as power sources, and spares for all of them must be carried, as part of the Soldier load, to meet the nominal dismounted operation time requirement of 72 hours.

Soldier Load

The poorly designed "everything on the Soldier" approach to support dismounted operations significantly stresses the Soldier and is the largest contributor to noncombat injuries: 24 percent of medical evacuations in OIF and OEF have been for non-combat musculoskeletal injuries.[2]

FIGURE D-1 Soldier with combat load. SOURCE: Dr. Marilyn Freeman, Deputy Assistant Secretary of the Army for Research and Technology, "Providing Technology Enabled Capabilities to Soldiers and Tactical Small Units," presentation at the 2011 AUSA ILW Winter Symposium and Exposition, Fort Lauderdale, Florida, February 23, 2011.

Force Protection

Force protection measures to ensure the highest degree of survivability are uneven across the spectrum of operations performed. Body armor focuses on protection of the torso and head, and its significant weight increases the Soldier's exposure to harm and contributes to the Soldier load problems.

Unregulated Fielding of New Technology

On multiple occasions, committee members heard from military combat veterans about technology "solutions" that had been rapidly fielded to the OIF/OEF theater of operations but

[2]COL Gaston P. Bathalon, Commander, Army Research Institute of Environmental Medicine, U.S, Army Medical Research and Materiel Command, "The Soldier as a Decisive Weapon: USAMRMC Soldier Focused Research," presentation to the Board on Army Science and Technology, February 15, 2011.

were never used in TSU operations. Instead, the new technologies ended up being stored in a CONEX (metal shipping container) for various reasons, including human-system interface problems; lack of training; excessive weight for value added; and lack of integration with existing systems.

Deficits in Soldier Resiliency

Recent successes with "resilience-training" programs indicate that resiliency has been a problem, and much more improvement is needed. As evidence, there were 303 suicides in calendar year 2010, which is about double the number in 2003. After returning from deployments, 20-40 percent of Soldiers had been referred for mental health problems such as traumatic stress disorder, depression, and interpersonal conflict.[3]

The Army has released two important reports on the health risks, including behavioral health and risks such as suicide and prescription drug abuse, faced by the active force and veterans of OEF and OIF.

- *Army Health Promotion, Risk Reduction, Suicide Prevention Report 2010* reported on "indicators of stress on the force and an increasing propensity for Soldiers to engage in high risk behavior." In addition to the 239 suicide deaths across the entire Army (including the Reserve component) in FY 2009, 160 of whom were active duty Soldiers, the report noted that there were 1,713 known suicide attempts during that same period (U.S. Army, 2010, p. i).
- *Army 2020: Generating Health & Discipline in the Force Ahead of the Strategic Reset: Report 2012* documents and emphasizes the interrelatedness of health and disciplinary issues ranging from posttraumatic stress and other behavioral health disorders to illicit drug use, other high-risk behaviors, and suicide (U.S. Army, 2012). For its update of the Health and Disciplinary Maze Model, which had been introduced in the 2010 report, the FY 2011 statistics included more than 42,000 criminal offenses, of which more than 11,000 were drug- or alcohol-related, as well as 1,012 known suicide attempts and 162 suicides (U.S. Army, 2012, p. 6). Newspaper accounts of the personal tragedies for Soldiers and their families, such as an April 2012 op-ed column in the *New York Times* (Kristof, 2012), help to put human faces on these awful statistics and given them real-life meaning.

A policy brief from the Center for a New American Security states that, from 2005 through 2010, service members across all branches took their own lives at an average of one death every 36 hours (Harrell and Berglass, 2011). Army suicides have climbed steadily since 2004, while suicides in the Air Force, Navy (other than Marine Corps), and Coast Guard have been stable. Although accurate data on veteran suicides are not available, the Veterans Administration estimates that a veteran dies by suicide every 80 minutes (Harrell and Berglass, 2011). Like the Army reports, this policy brief notes that risk factors for suicide include

---

[3]COL Gaston P. Bathalon, Commander, Army Research Institute of Environmental Medicine, U.S, Army Medical Research and Materiel Command, "The Soldier as a Decisive Weapon: USAMRMC Soldier Focused Research," presentation to the Board on Army Science and Technology, February 15, 2011.

traumatic brain injury and a range of behavioral symptoms associated with posttraumatic stress disorder.

Soldier Fatigue and Nutrition

Since 1992, more than 24,000 Soldiers have been discharged for failing to meet Army Weight Control Program requirements, and 20 percent of combat Soldiers suffer weight loss of more than 5 percent and performance deficits due to unmet nutritional requirements. Factors known to contribute to physiological and mental fatigue include night work, disturbed or restricted sleep cycles, rapid deployment across multiple time zones, and rapid deployment to significantly higher altitudes. TSU leaders appear to lack the training to ensure that their Soldiers receive the rest and nutrition they need to sustain high performance under demanding environmental conditions during challenging missions. [4]

[4]COL Gaston P. Bathalon, Commander, Army Research Institute of Environmental Medicine, U.S, Army Medical Research and Materiel Command, "The Soldier as a Decisive Weapon: USAMRMC Soldier Focused Research," presentation to the Board on Army Science and Technology, February 15, 2011.

REFERENCES

ASB (Army Science Board). 1991. Army Science Board 1991 Summer Study - Soldier as a System. Washington, D.C.: U.S. Department of the Army.

Harrell, M.C., and N. Berglass. 2011. Losing the Battle: The Challenge of Military Suicide. Available online http://www.cnas.org/files/documents/publications/CNAS_LosingTheBattle_HarrellBerglass_0.pdf Accessed March 28, 2013.

Kristof, N. 2012. A veteran's death, the nation's shame. Available online http://www.nytimes.com/2012/04/15/opinion/sunday/kristof-a-veterans-death-the-nations-shame.html?pagewanted=all. Accessed March 28, 2013.

U.S. Army. 2010. Army Health Promotion, Risk Reduction, Suicide Prevention Report 2010. Report of the Army Suicide Prevention Task Force. Arlington, Va.: Headquarters, Department of the Army.

U.S. Army. 2012. Army 2020: Generating Health & Discipline in the Force Ahead of the Strategic Reset: Report 2012. Arlington, Va.: Headquarters, Department of the Army.

Appendix E

Measures of Performance and Measures of Effectiveness

A detailed methodology is used by the Army's analysis community to develop measures of performance (MOPs) and measures of effectiveness (MOEs), especially those needed to make design trades among alternatives during design and development and to determine the relative contributions of multiple factors.

DEVELOPMENT SEQUENCE

The development of MOPs and MOEs follows a sequence of steps similar to the following:

- Identify a military utility (e.g., enhanced tactical small unit (TSU) effectiveness in stability operations) that can be impacted by a Doctrine, Organization, Training, Materiel, Leadership and Education, Personnel and Facilities (DOTMLPF) effort (e.g., access to local sociocognitive networks).
- With this military utility in mind, identify supporting objectives (e.g., determine access to sociocognitive databases).
- Once objectives are formulated, identify essential elements of analysis (EEAs), which are basically the key questions one might ask to support the objectives. For example, an EEA might be, What role does information exchange across a network play in the utilization of these sociocognitive databases?
- Identify issues that are derived from the EEAs. For example, information exchange (especially for digital images and streaming video) is very poor at the TSU level. Bandwidth rate is one issue. Another is that operations tempo does not give TSUs enough time to download, evaluate, and make judgments based on available information— that is it is very easy to reach information overload.
- Identify hypotheses that address each issue. For example, Soldiers and TSUs would benefit from advancements in dynamic communications, information, and sociocognitive networks for enhancements of information exchange and assessment of information.
- Identify the data needed to prove or disprove each hypothesis.

- Identify metrics (the MOPs and/or MOEs) needed to collect the needed data.
- Develop scenarios that will generate opportunities for the collection of data to measure performance and effectiveness.

In conducting assessments for the dismounted TSU and Soldier, the Army should use a methodology similar to that outlined above to create the most appropriate system and system-of-systems metrics—MOPs and MOEs. The MOPs should assess what Soldiers or TSUs achieve in terms of technical performance. In general, MOPs used by the Army are quantitative, but they can also apply qualitative attributes to task accomplishment. Simply put, MOPs measure what Soldiers and TSUs are doing but encourage the system designer and evaluator to ask whether the TSU or the Soldier is doing the right things to achieve the desired effect. Examples of Soldier MOPs include measurable enhancements to Soldier mobility and endurance (e.g., due to offloading physical and mental loads, enhancing nutrition, improving sleep cycles, and altering mission duration times); measures of ability to develop Level I situational awareness; ability to be "culturally correct" when interacting with local nationals; reductions in the probability of being hit by threat munitions because of improvements in agility; and assessments of the Soldier's sensory (visual, auditory, tactile, and olfactory) perception using measures such as detection, position, recognition, identification, time, distance, error, etc. Examples of TSU MOPs include measures of ability to integrate nonorganic fires and effects, ability of TSU to shoot down incoming unmanned aerial threats (e.g., small drones), and the ability of the squad to offload and then recover equipment before and after a mission, as well as the time needed for TSU leaders to accurately convey appropriate parts of a mission plan (or fragmentation order) to all members of the TSU, and the time needed to achieve a mission.

MOEs assess the impact of the actions of the TSU and the individual Soldier on the effectiveness of achieving mission and task objectives. These measures assess changes in behavior, capability, or operational environment; they do not measure task performance. They measure what is accomplished and help to verify whether objectives, goals and end states are being met. They are typically more subjective than MOPs and can be defined as either qualitative or quantitative measures. For instance, an MOE may be based on quantitative measures to reflect a trend and show progress toward a measurable threshold. Examples of Soldier MOEs include the percentage of time a Soldier is distracted from focusing on the mission/objective, measures of the ability of a Soldier to exploit his situational understanding, and measures of the ability of a Soldier to contribute to TSU effectiveness. Examples of TSU MOEs include measures of ability to engage enemy threats outside the range of enemy weapons, ability to successfully achieve the commanders intent, percentage of time the TSU is surprised by the enemy, ability of the squad to rapidly adapt (mentally and physically) to loss of personnel or a warfighting capability, ability to enhance individual Soldier Level II and

Level III situational awareness through shared situational awareness, and timeliness and ability of the TSU to react to significant enemy actions.

Indicators

In addition to MOEs and MOPs, a systems engineering approach will also require appropriate indicators. An indicator is an event that serves as evidence that an effect is being accomplished or, for an MOP, that an output outcome is being achieved. Good indicators are clear, concise, and, most important, reasonably related to an MOP or MOE. Indicators may be quantitative (e.g., number of weapons needed and/or shots needed to shoot down a drone) or they may be qualitative (e.g., number of subject matter experts who agree that a TSU achieved the commander's intent). A single indicator can support more than one MOP or MOE. For example, a reduction in the number and length of radio calls within a TSU may be an indicator that there is better shared situational awareness (an MOE); a significant positive impact of Soldier, TSU, and leadership training methodologies (an MOE); more time for the Soldier and TSU to focus on assigned tasks and missions (an MOE); enhanced individual cognitive performance (an MOP); a well-designed Soldier-centric interface to the TSU network (an MOP); and a high-performing information-sharing system on the TSU network (an MOP).

Why MOPs and MOEs Are Important

The lack of adequate MOPs and MOEs has brought a lack of accountability for dismounted TSU and Soldier performance. Perhaps more important for the subject of this report, the lack of MOPs and MOEs that realistically assess both human and materiel contributions to required capabilities has vitiated real progress toward holistic design and evaluation of the TSU and the Soldier, despite a history of advice on achieving that end (see Appendix D).

Compared to the Marine Corps, the Army light infantry squad has had an unstable organizational structure and size. No one at the Infantry School or in the research and development (R&D) centers could give the committee a rationale for the current nine-person size of the dismounted TSU other than military judgment. In fact, other infantry-like formations in both the Army (e.g., Special Forces) and other military Services have explored alternative squad sizes and structures, and there seems to be no clear consensus that the current squad size is optimal for any specific environment, let alone all environments encompassed by unified land operations.

At the Army Maneuver Center of Excellence and at the R&D centers visited by the committee, many training technology demonstrations were briefed, but few had been widely adopted. Comments from the two roundtables with postdeployment noncommissioned officers (NCOs) suggest these combat-proven TSU leaders were unaware of many of these training technologies and were too rushed in their deployment readiness training to make use of new training opportunities or approaches.[1] The committee found no evidence of objective metrics to indicate state of training for units preparing to deploy or in theater.

Communications, intelligence, and logistics technologies briefed at the R&D centers visited by the committee as having been demonstrated with positive results and available for more widespread fielding or use were described in the NCO roundtables as "Conex-fillers," which were too much trouble to learn how to use and exploit. Although some of the NCOs attributed these lost technology opportunities to "drive-by fielding," the committee believes a more likely explanation is the lack of appropriate tactics, technique and procedures to guide their use, of system integration, and of training resources to enable TSU mastery of the available technology prior to deployment. Accountability to TSU performance metrics would be an incentive for TSU leaders to continually seek better approaches, including new technologies.

Substantial knowledge exists about the relationships between nutrition and physical and cognitive performance directly pertinent to TSU performance. Medical and food technology scientists at the Natick Soldier Research, Development, and Engineering Center reported that available rations were underused or misused. Indeed, based on the committee's observations during site visits, this appears to be a feature of infantry training. Early-stage Soldiers learn informally during training how to take apart or "field strip" the carefully constructed and designed rations now being deployed. Leaders trained to see the relation between what their Soldiers are or are not consuming and trainers teaching Soldiers how to eat could improve overall performance and endurance and make better use of the rations provided. Furthermore, having MOPs and MOEs for field performance will provide baseline performance levels from which to evaluate potentially useful new developments.

MULTIVARIATE ANALYSIS OF MOP AND MOE DATA

To adequately assess the military performance and effectiveness of the Soldier and the TSU as a complex system or system of systems encompassing all the DOTMLPF domains, a systems analysis effort such as multivariate analysis is needed that can support detailed estimation and prediction techniques. A multivariate analysis involves the observation and analysis of multiple variables at the same time. From a systems engineering perspective, this type of analysis is used to perform trade studies across multiple dimensions while taking into account the effects of all variables on the military performance or effectiveness being assessed. Variables are identified as dependent (that which is being

[1]Informal discussions between the Committee, and noncommissioned officers and officer candidates, during Meeting 3, July 12, 2012, in Ft. Benning, Georgia, and between the Board on Army Science and Technology and noncommissioned officers and officer candidates, February 23, 2010, in Fort Bliss, Texas.

measured) and independent (the DOTMLPF characteristic being manipulated or changed).

Depending on the design of the experiment, a variable may be dependent in one case and independent in another. If the amount of information flowing to an individual is varied, one may observe an increase in the cognitive workload (the workload may go up with too much information; it may also go up with too little or no information). In this case, the cognitive workload is the dependent variable, which is a function of the amount of information (independent variable) that is presented to the Soldier. In this case, the cognitive workload could be used as a level of performance.

Likewise, if cognitive workload is varied, one may observe a variability in the quality (e.g., in terms of appropriateness or timeliness) of decisions being made. In this case, the quality of decisions (dependent variable) is a function of the cognitive workload (independent variable) and the quality of the decisions is a measure of effectiveness. Note also that "appropriateness" of decisions may be a subjective assessment that needs "indicators"—for example, blue forces unnecessarily sent in harm's way, choice of approach route that does not offer the tactical advantage of other routes, calls missed from subordinates and supervisors) to validate its assessment.

Appendix F

Simulation Technologies and Devices

There are a number of training simulations and devices currently fielded that go a long way to meeting the capability needs of tactical small units. This appendix highlights several areas that could accelerate and accentuate the effectiveness of training through simulation technologies that span the live, virtual, and constructive training spectrum.

AUTHORING TOOLS

There are now established protocols for performing a cognitive task analysis (Clark et al., 2008). The result of this process is a set of learning objectives and knowledge that can be used directly in a guided experiential learning system. What is needed is a set of authoring tools to guide training developers in the steps for creating a training package using experiential and realistic live, virtual, and constructive simulations that supports cognitive task analysis and instructional design. The authoring tools should enable the authoring and linking of learning objectives, assessment, and feedback for coaching in the context of an experiential training scenario.

Authoring tools can also be used to support rapid, complex scenario development. A given issue that has often been raised is the difficulty of authoring complex scenarios for training. Scenario authoring currently takes a long time and a great deal of expertise. Given a content library and a virtual environment or game engine, authoring tools should enable training developers to rapidly design a new scenario or edit an existing scenario in a fraction of the time currently needed. To the extent possible, the authoring tools should make use of real-world scenario data, but the scenario should relate to the learning objectives produced during the instructional design.

Authoring tools can also improve the content creation pipeline to support the virtual worlds used in games and simulations for training. One of the greatest expenses in development of games and simulations is the creation of the artwork and animation to bring about the desired learning effects. The cost and time of training development can potentially be reduced by an order of magnitude by content development pipelines that automatically acquire, model, and animate objects and people, reducing the need for support by an artist. While commercial game developers and special-effects houses are developing tools to make it less expensive to rapidly develop content, they typically do not make these tools available to others due to the competitive nature of the marketplace.

TOOLS FOR IMMERSION

The need to immerse the trainee in a virtual world for more effective learning, while at the same time keeping the cost of the display system low, is an important issue. It is becoming increasingly possible to develop low-cost, small-footprint, ruggedized, wireless, head-mounted display systems to provide the learner with a realistic visual experience in the virtual world. Further, it would be ideal to improve head-mounted displays by making them fit the size and form factor of the soldier's eye protection wear.

The other alternative worthy of further investigation is to bring the virtual world out to the physical world. This concept was demonstrated in the Future Immersive Training Environment Joint Capabilities Technology Demonstration at the Infantry Immersion Trainer (IIT) at Camp Pendleton. The IIT used display screens that featured interactive characters integrated into a physical site used in training for military operations in urban terrain. An alternative to the display screen technology is the head-mounted projectors with retroreflective material, which gives each viewer an individualized view of the world. Training immersion can also be achieved through naturalistic interfaces to computer systems and autonomous animated nonplayer characters and teams.

Naturalistic Interfaces

For dismounted transport in games and simulations, the current standard interface is a game controller or a joystick, which is not a natural way of moving or using one's body in the virtual world. The training technology community should leverage the trend toward vision-based interfaces that track the body and facial expressions and infer gestural meaning. In addition the community needs to leverage advances in speech recognition and natural language processing to enable conversational interfaces to games and simulations.

Autonomous NonPlayer Characters

One of the limitations of using platforms like VBS2 and other game-based simulations is that the avatars of the opposing forces and civilians have to be controlled by other humans. Like the Janus simulator, which required six people to train one person, this is a costly way to do business. While many of these systems have semiautonomous forces, their capabilities still require supervision by human exercise controllers. An alternative is to develop programmable autonomous characters and teams that will play the roles of the opposing force and civilians across all the mission sets: offense, defense, and wide-area support operations in the Dismounted Soldier Training System. Autonomous characters should be capable of perceiving objects and entities in the virtual environment, making plans and decisions, taking coordinated action with or against human teams, and providing feedback to the after-action reporting system. They should be capable of being used in game-based environments such as VBS2 and the CryEngine as well as in a Massively Multiplayer Online Game environment such as EDGE, which is currently being developed by the Army Research, Development and Engineering Command. They should also be capable of explaining their actions and decisions during after action review so that squad members can see how their actions affected the opposing force.

REFERENCE

Clark, R.E., D. Feldon, J.J. Merrienboer, K. Yates, and S. Early. 2008. Cognitive Task Analysis. Pp. 577-594 in Handbook for Research for Educational Communications and Technology, edited by J.M. Spector, M.D. Merrill, J.J. G. van Merriënboer, and M.P. Driscoll. Mahwah, N.J.: Lawrence Erlbaum Associates.

Appendix G

Technology Solutions for TSU Sensor Missions

Sensor and optics technology is employed to increase the effectiveness of the squad. A sensor system transduces propagating energy encoded with certain information on the threat environment into a format usable by the Soldier. Examples include full motion video (FMV), infrared search and track (IRST), radar, communications intelligence receivers, acoustic unmanned ground sensors, and acoustic sniper detection systems. Generally, the waveband determines the type of sensor and its potential utility.

Squad-level sensors are used in three types of missions:

- Situational awareness (SA)[1],
- Force protection; and
- Precision targeting.

Other sensor missions—most notably intelligence, surveillance, and reconnaissance (ISR)—support the squad but are generally accomplished at higher levels or by other organizations. Sensors providing situational awareness yield timely information about current events in the space around the squad, such as the locations of dismounted threats, approaching vehicles, or potential targets within buildings. Force protection sensor technology focuses on providing adequate warning to minimize lethal engagements involving rockets, artillery, mortars, small arms fire, mines, improvised explosive devices, and chemical-biological-radioactive-nuclear (CBRN) agents. Precision targeting sensors provide highly lethal fire-control information to blue force weapons; examples include infrared seekers or the counter-battery solution generated from weapons location radar (WLR). Navigation sensors applied in a GPS-denied environment is a squad-level consideration falling under the situational awareness umbrella. Electronic warfare is a very important consideration; however, for the purposes of this discussion, is only considered where there is electronic protection (anti-jamming).

The following are some guidelines for employing sensors for small-unit ISR:

[1]The SA sensor mission is to gather information that can be used to increase Soldier/TSU situational awareness.

- Sensor technology should create a decisive advantage and in no way impair other tasks critical to the squad's operations. The tasking, collection, processing, exploitation and dissemination (TCPED) process for the technology should be carefully developed with full attention to human-system interface factors.
- Sensor technology should be interoperable and modular.
- Soldiers should train with available sensor technology. Understanding the capabilities and limitations of available technology is critical.
- Unmanned sensors (unattended ground sensors (UGSs), unmanned aerial vehicles (UAVs), micro air vehicles (MAVs), etc.) can be used to extend the range and influence of the squad. An unprecedented degree of autonomous platform and sensor operation must be created.
- Given advances in network integration and small arms lethal ranges, a small unit leader must have the capability to "see" movement to 1,800 meters in all terrain and the ability to determine the character of that movement (armed men or civilians) out to 900 meters.[2]
- Limiting the local range of operation ensures use of sensors with appropriate space, weight, and power (SWAP) and constrains information presented to the squad to that which is most pertinent. Special consideration should be given to the sensor architecture and operating environment. For example, can the sensor employ cell tower transmissions as a source of illumination for moving target indication?
- Sensor technology should be developed in full consideration of the roles and responsibilities of every member of the squad.
- Distributing capability among squad members would be prudent.

Sensors may be either passive or active. Because passive sensors do not require their own sources of illumination they tend to be more stealthy than their active counterparts. Thermal imagers and signals intelligence (SIGINT) receivers are examples of passive sensing modes. Active modes employ transmitters to propagate energy; this transmitted energy convolves with the target impulse response to yield a target signal at the receiver. Radar and laser rangefinding are examples of active sensor systems. Generally, passive systems are lower cost, lighter weight, and use significantly less power to operate.

SWAP and cost (SWAP-C) are the main constraints on materiel solutions and performance achievable by a given sensor technology. For example, weapons location radar may be used at a forward operating base, but for practical reasons (size, weight, power, and deployment) will not accompany a squad during a typical engagement. It is important to consider the array of squad missions and

[2]Surveilling a 900m-1800m ring is nontrivial with optics. Rather, low power radar and SIGINT can be used for SA information and then can cue EO/IR for target characterization. An approach for meeting these needs is a network of sensors. Sensors organic to the tactical small unit (TSU) could meet the short range requirements, while networked access to supporting sensor systems could satisfy the longer range needs.

determine where enhanced sensor capability can provide a decisive edge and the practicality of deploying that capability given the squad's SWAP-C constraints.

Information gained from sensor technology is often achieved through sophisticated, automated processing. Different types of information come from the different sensor missions. Situational awareness may require sensor outputs that yield threat location and estimated characteristics (such as range rate, predicted positions, and types of targets). Force protection systems generate a warning response when a threat is deemed present and may then direct an interceptor; threat interception, at the very least, requires some knowledge of threat bearing and may otherwise need target class and state estimates for complete tracking. The ability to seamlessly and effectively combine multisource data into a common picture is advantageous for enhanced situational awareness and force protection. Sensors supporting precision targeting must provide information of sufficient timeliness and quality to meet weapon requirements; typical operator interaction involves prioritizing and confirming targets.

Managing sensor-generated information is the biggest challenge facing the squad. Unlike ISR data products, which may occupy multiple intelligence analysts and function with a latency of minutes or hours, the squad requires actionable information with delays of a few seconds or less. Achieving the goal of tasking sensors, collecting necessary data, processing and exploiting the data, and disseminating important information—including, potentially, fusing data from other sources, such as National Technical Means—is an important challenge in providing the squad with superior SA and an overwhelming advantage over the adversary. (Fusing data from ISR sources generally requires multilevel security to protect data collection means and sensor features.) Tasking and collection is a time-consuming operation and has to be automated to minimize the impact on the squad's mission.

Materiel solutions supporting the squad should be based on an open system architecture (OSA). An OSA enables the integration of different sensor technologies and products from different vendors by defining system interfaces and data formats. This modular, open approach has a number of critical benefits, notably the potential for significantly reduced SWAP-C, a means to tailor sensor packages for different missions and target types, reduced learning curve and training requirements, and the ability to share time-critical information in straightforward fashion. The OSA should accommodate multilevel security considerations.

SPECIFIC SQUAD-LEVEL SENSOR CONSIDERATIONS

Table G-1 highlights key sensor considerations. Sensor requirements are specifically tied to mission objectives. In general, physical considerations—size, weight, power, deployment platform—impact the details of the sensor design. From the squad's perspective, the sensor must be easily maneuvered. For example, a through-the-wall radar, used to detect activity or find weapons caches in a building, must be lightweight and ideally weapon-mounted so that the Soldier is always able to fight while in threat areas. Moreover, powering the sensor should not overly burden the squad with battery and sensor mount weight.

It may be that additional SWAP is warranted if the sensor provides an overwhelming advantage for a designated mission (e.g., finding weapons caches behind walls). This requires sensors with well-defined performance characteristics—such as detecting heartbeats or finding weapons caches with a specified probability—and robustness across the range of practical operating environments. Sensors must cope with a range of environmental characteristics, including urban clutter, multipath, different building characteristics, background traffic, mountain clutter, varying interference characteristics, etc. Additionally, the concept of operations (CONOPS) may affect sensor performance (e.g., in a through-the-wall sensor, the system must ensure it is detecting the threat heartbeat and not the operator heartbeat). Processing complexity is one approach to ensure increasingly robust performance, but embedded computing using modern sensor processing algorithms generally has high SWAP requirements.

TCPED is the process that makes sensor technology useful to the warfighter. In many instances, whole communities are available to support TCPED. The enormity of the approach suggests an array of sophisticated sensors and well-trained analysts using complicated tools to derive critical information. Naturally, this traditional approach is of significantly less value to the squad. Rather, TCPED for the squad requires an unprecedented degree of automation and very low latency. Sensors must be positioned and tasked to collect data over very short time intervals. This data must then be processed, exploited, and disseminated to the squad to help guide tactical decision-making. Automation is the only practical way to close the TCPED loop and ensure that sensor technology does not adversely preoccupy the squad.

The human-system interface (HSI) is a critical sensor design consideration. Anything that affects the squad's cognition requires serious evaluation. The sensor TCPED process should provide information to the squad that is of high value, timely, and actionable. Providing this information in a useful manner that enhances the squad and does not detract from any other basic functions—such as performance in a firefight, interacting with locals, working as a team, etc.—is an intrinsic requirement. Additionally, the Soldier should understand the utility and quality of the sensor data, lest important information be discarded and less useful information be acted upon. Developing a test range to explicitly support acquisition and deployment of sensor technology and at the same time ensuring the incorporation of HSI best practices into squad-level sensor design, is imperative.

An OSA employs specific system interfaces (e.g., inputs, outputs, and power usage) to ensure interoperability of subsystems and an integrated approach to sensor design. Key advantages of OSA standards from the squad's perspective include modularity to support an array of missions; simplicity in integrating new technologies; support for a common user interface in abidance with HSI design goals; a means to matrix sensor capabilities across the squad; a mechanism to acquire and integrate technology from a broader vendor base; and a common framework to train the squad on sensor technology usage. A government-owned OSA is necessary to ensure adherence to the spirit of the approach: Multiple vendors can participate and help enhance the OSA, with emphasis on sensor characteristics, performance, and utility.

TABLE G-1 Squad-Level Sensor Considerations

Issue	Comments	Key Considerations
SWAP	Limiting operating range and field of view will minimize SWAP. Deploying on autonomous vehicles ideal for many missions, but may cue adversary to squad presence.	Modular, open approach tied to strong systems engineering and detailed training. Approach seamlessly tied to CONOPS.
Performance	Detection performance, false alarm rate, location accuracy, target parameter estimation accuracy, resolution and contrast must be tied to squad's specific mission goals. May vary by mission.	Operations in complex clutter and interference environments. Impact of multipath and operator/platform motion. Computing processor power and overall SWAP. Minimizing impact on Soldier's cognition.
TCPED	Effective management of sensor assets requires a hands free TCPED process for the squad.	Focus on regions around the squad, tie in ISR data from other sources with appropriate confidence weighting.
Human-system interface	Data must be presented to squad in most effective and primitive form. No time or resources for interpretation. Example: blue dots for incoming threats, red dots for fleeing threats ("blue is new, red has fled")	HSI, coupled with TCPED, is the single most important consideration in providing the squad the best and most useful sensor technology. Training should be included under HSI.
Open system/modularity / cost	Allowing multiple vendors to compete for support systems, thereby enabling significantly less costly subsystems of product improvement. Interoperability of subsystems is a key feature.	The government should consider developing and owning the OSA.

Issue	Comments	Key Considerations
Multilevel security	Integration of information from multiple sources owing to security constraints is a challenge.	Well-defined metadata formats integrated into the OSA allow key information to be provided to the squad without divulging sensor characteristics and sources and means.
Systems engineering	Methodical approach to developing system requirements and understand relationships among systems of systems in the engineering process.	Couple Soldier characteristics and training with physics-based models of sensor capabilities. Develop systems engineering experience within the government team. Focus on OSA as key approach to interoperability and a modular approach to building squad capability.

Multilevel security is a known impediment to timely and broad dissemination of information to the squad. It is highly unlikely that anyone in the squad will have security access to the wealth of information gathered by DoD and intelligence community sensors. Moreover, much of the data from these other sources is of a strategic nature: it provides important context but may not possess the timeliness of the information required by the squad. The ability to incorporate data from other sources to help manage sensor information collected by squad-level assets would necessitate a mechanism to downgrade security. An effectively designed OSA is able to support this objective by separating critical sensor information from those items that characterize a sensor's physical design or sources and means of data collection. In the context of the OSA, metadata formats that convey the threat details of most interest to the squad (e.g., regions where the threat was last observed and general threat characteristics) is an effective approach to interface systems of varying classification.

While systems engineering may have different meanings, in the context of building sensor technology for the decisive squad of the future it points out the process to specify sensor requirements. These requirements comprise all critical considerations, including performance, subsystem interoperability, SWAP, HSI, Soldier training, and life-cycle management. Effective systems engineering requires highly competent and well-trained acquisition professionals and support infrastructure, as well as effective software tools, test ranges, and acquisition strategy. From a strategy perspective, enforcing specific acquisition standards (e.g., OSA compliance), efficiently framing system requirements, and shortening the acquisition cycle to enhance the cost-risk-benefit trade space all appear essential in better supporting the squad's equipment needs.

Situational Awareness

Situational awareness provides the squad with current information on the threat environment. Threats include dismounts approaching the squad or leaving the vicinity, vehicles in proximity to the squad, CBRN and explosive emplacements, weapons caches, and blockaded routes. Dismounts may be combatants or the general populace, in the open or behind cover, and carrying weapons or unarmed. The complexities of the operational environment make gathering SA a challenge: clutter environments can appear extremely heterogeneous; multiple, potential targets may occupy the search space; complete SA may require propagation through anisotropic media (e.g., layered or mixed building materials with air gaps); urban or mountainous settings can create severe multipath scenarios and obscuration; users inadvertently interact with the sensor, or the sensor platform requires specialized motion compensation techniques; and intentional or unintentional interference degrades sensor performance.

From the squad's perspective, the ability to provide SA for a diameter of 900 meters centered on the squad is highly desirable. Limiting the SA window minimizes the amount of information deluging the squad. Providing the right information is critical. SA may be divided into a secondary sector of interest, perhaps out to 1,800 meters with focus on specific threats (e.g., vehicles only) that might enter the 900 meter inner region. In addition to the typical ground threats engaging the squad, future threats may include unmanned aerial systems (UASs), helicopters, and other small aircraft. Intelligence forecasts of the threat environment are essential in the SA sensor acquisition process.

Anticipated threats where superior SA will greatly enhance the squad's performance include these:

- Dismounts;
- Ground vehicles, including trucks, cars, and motorcycles;
- Obscured targets (dismounts, weapons, and weapons caches) within buildings or natural structures or under foliage;
- Concealed weapons carried on dismounts; and
- Detection, characterization, and location of emitters.

Additional SA missions include navigation in GPS-denied environments and life signs monitoring. SA against small airborne threats, specifically UASs, may prove an important mission in the near future. Table G-2 summarizes the SA sensor missions.

Dismount and vehicle detection can involve radar, FMV, and IRST sensors. Radar has the widest field of view and is generally preferred for larger area search and quicker responses. Additionally, radar encodes target presence and motion on the amplitude and phase variation of the radiofrequency signal; automated radar signal processing methods to detect targets in strong clutter and interference environments continue to show significant advancement with corresponding improvements in sensor (e.g., multichannel arrays and waveform agility) and computing technology. FMV has a generally limited field of view and thus has limitations in a search mode. Recent developments extend FMV capability to wider areas by using multiple telescopes and processing to stitch the resulting outputs into a common picture; this technology is called wide area motion imagery (WAMI). Moving targets are automatically tracked in FMV and WAMI by finding and tracking pixel changes from frame to frame. Similarly, IRST sensor technology searches for regions of high emissivity to detect targets and then observes changes in pixel emissivity from frame to frame to estimate vehicle motion.

TABLE G-2 Squad-Level Sensor Missions

Mission	Description	Objective	Relevant Sensor Technology
SA	Dismount detection and engagement	Detect, locate, characterize, and monitor dismounts in vicinity of squad	Radar, SIGINT, FMV, IRST, WAMI
SA	Vehicle detection and engagement	Detect, locate, characterize, and monitor vehicles in vicinity of squad	Radar, FMV, IRST, WAMI, acoustics, seismometer
SA	Through-wall surveillance	Determine presence of possible combatants and weapons caches within buildings and structures	Radar, SIGINT
SA	Foliage obscured target surveillance	Determine presence of possible combatants, weapons, and weapons caches under foliage	Radar, LAser Detection And Ranging (LADAR), SIGINT
SA	Spectrum surveillance	Find and characterize emitters within vicinity of squad	SIGINT
SA	Navigation in GPS-denied environments	Provide position information in absence of traditional, handheld GPS device	Radiofrequency sensor technology

Mission	Description	Objective	Relevant Sensor Technology
SA, force protection	Concealed weapons detection	Detect concealed weapons among general populace	Millimeter wave radar, metal detector, magnetometer
SA, force protection	Life signs monitoring	Remote detection of fallen-Soldier life status	Millimeter wave radar, acoustics, laser
Force protection	Perimeter surveillance	Force protection in vicinity of encampment	Radar, IRST, acoustics, SIGINT
Force protection, precision targeting	Counterrocket, artillery, and mortar (CRAM)	Force protection in vicinity of encampment	Radar, IRST, acoustics
Force protection	Counter-improvised explosive device (CIED)	Detect and locate likely improvised explosive device (IED) emplacements	Radar, HSI/MSI, SIGINT
Force protection, precision targeting	Counter-shot/sniper	Detect location of small arms fire	Acoustics, IRST
Force protection	Mine detection	Detect buried mines	Ground penetrating radar, HSI/MSI, magnetometer, metal detector
Force protection	CBRN agent detection	Detect threatening agents to support evasive actions	CBRN-tailored sensors, remote sensing techniques (radar, EO/IR, HSI/MSI)
Precision Targeting	Vehicle engagement	Target armored and nonarmored vehicles	IR, radar, optical sights

Mission	Description	Objective	Relevant Sensor Technology
Precision targeting	Concealed Threat Targeting	Engage threats hidden behind abutments, on other side of buildings, etc.	EO/IR, radar imaging

Note: HSI/MSI, hyperspectral imaging/multispectral imaging.

Obscured targets include objects within buildings, under foliage, or buried under soil. From an SA perspective, identifying all threats close to the squad is the goal. Lower frequency radar—about 1 GHz and below—is the technology of choice for obscured target detection. These lower frequency signals penetrate many building materials and foliage. Signal attenuation is severe and multipath can prove problematic, thereby limiting the system operating range. Laser detection and ranging (or LADAR, sometimes called light detection and ranging or LIDAR) is also a useful technology to map detected threats under foliage; LADAR generally does not detect sense-through-the-wall collections but can be used against targets under foliage when the laser has line of sight to the target.

Millimeter wave (mmw) radar can detect concealed weapons at moderate ranges. Such radars (typically in the range 35-95 GHz) are generally smaller systems of comparable or better resolution than their lower frequency counterparts. New airport surveillance technology uses mmw scanners at security checkpoints. Squad applications would likely be for concealed weapons detection at ranges of a few meters to tens of meters. Magnetometers and metal detectors are not likely to have application, since they operate over shorter distances than mmw radar.

Detection, characterization, and location of enemy emitters is a SIGINT function. Typical emitter threats are in the radiofrequency range, though SIGINT receivers have been developed to intercept laser-modulated signals. SA against all common emitters—typically, cell phones in the GSM bands and other handheld radios down to VHF—is of value to the squad. The common approaches to SIGINT collection include the use of a single, multiaperture receiver with sufficiently long baseline to achieve accuracy goals; multi-sensor intercept topologies (fusion of the intercepts from two or more sensors); or, moving a single intercept receiver through large integration angle and using frequency or time-differencing techniques. This latter approach requires greater time and may have limited applicability to the squad's SA needs. SIGINT sensors can apply to obscured target detection when an emitter is present, such as a cell phone or a key fob or other exploitable device; lower frequency operation is essential to minimize signal attenuation through the obscuring medium, but SIGINT incurs only one-way loss (as opposed to two-way loss in radar).

Navigation in a GPS-denied environment can employ coded radio frequency signals and multilateration in a local network, using principles similar to those used by radio navigation satellite systems, such as GPS, Global Navigation Satellite System, Galileo, and the Computerized Movement Planning and Status System. Generally, four transmitter sources are needed to determine location in three dimensions and account for time, yielding absolute position. Frequency diversity is also required to minimize multipath effects on geolocation performance.

Life signs monitoring is considered separate from through-the-wall target monitoring. A sensor is used to determine the life and health status of a fallen comrade. This technology can help protect members of the squad during a firefight or in other compromising situations. One approach is to use mmw radar to detect respiration. Depending on the range to the target, mmw radar can also be used to detect and calculate heart rate. LADAR detects respiration but not heart rate.

The above discussion suggests the need for multiple sensor assets operating over different frequency regimes. Identifying and developing multipurpose sensor packages (e.g., a single aperture to provide both radar and SIGINT capabilities; or, a single sensor for concealed weapons detection, life signs monitoring, and navigation) would be highly desirable.

Force Protection

The primary force protection objectives (Table G-2) include the following:

- Perimeter surveillance for encampments;
- Early warning for incoming rockets, artillery, and mortars;
- Counter improvised explosive device (CIED);
- Fire/sniper location; and
- CBRN and explosives detection.

Life signs monitoring and concealed weapons detection fall in both domains, SA and force protection.

Perimeter surveillance provides early warning of an attack on an encampment. Radar, acoustics, and infrared (IR) sensors are likely sensor technology choices. Radar and acoustic sensors search for Doppler-shifted returns indicative of motion in the vicinity of the encampment; dismount targets have a very specific radar and acoustic signature, a "whooping" sound, predominantly due to torso motion (the radar signal can be converted to an acoustic output, and this is sometimes done in perimeter surveillance radar systems).

Early warning against rockets, artillery, and mortars is commonly the domain of WLR systems. WLR systems first detect the incoming threat and provide a warning. Then they calculate a counterfire solution based on the threat type and trajectory. IRST sensors can also be used; however, the false alarm rate in occupied environments (e.g., urban environments with dense traffic backgrounds) is a bigger concern than it is for radar. Active protection systems calculate the presence of an incoming threat and then deploy a kinetic kill response; again, radar is best suited for this difficult volume search and incoming threat location challenge.

Counter-IED sensor systems generally seek to find the threat and any details about its emplacement. In this sense, counter-IED benefits from ISR system products, such as change detection outputs and subsidence measurements. From the squad perspective, however, some capability for rapid determination of likely IED emplacements is essential. For example, a sensor system to scan and determine whether cordoned culverts have been tampered with would prove invaluable to a squad. Similarly, the ability to remotely scan and determine the presence of CBRN and mines is another key to force protection. A likely strategy for CBRN agent detection is to remotely interrogate sensors whose properties (e.g., radar cross section, resonance, luminescence) change when exposed to the target agent; radar, laser, and IR sensors are applicable technologies. Ground penetrating radar and hyper- or multispectral imagers can be used to detect disturbed soil and find buried mines at limited depths; such sensors exhibit variable performance depending on soil properties.

Rapidly determining the general location of hostile fire provides the squad with time to take cover and prepare to return fire. Of the technologies available to determine shot location, acoustic sensors appear best. IR sensors can detect muzzle flashes, but the search problem is challenging, and background clutter is a concern. The small radar cross section of a bullet renders radar less useful in this particular case. Multipath in urban and mountainous terrain is a limiting factor for acoustic-based hostile fire indication systems.

Precision Targeting

Squad-level precision targeting objectives include the following:

- Solutions against batteries of rockets, artillery, or mortars;
- Counterfire solutions against small arms;
- Vehicle engagement with high probability of kill; and
- Concealed threat targeting.

WLR systems employ target tracks and kinematic models to estimate the location of the hostile fire. This counterbattery solution is then used to return fire. Acoustic sensors are the likely choice to locate small arms fire; time-difference of arrival among several microphones, for example, can be used to determine the threat location. Vehicle engagement can involve fixed or moving targets detected by radar, FMV, WAMI, or IR sensors. Given likely constraints on squad engagement ranges, a multimode seeker fusing radar, laser, and IR sensors will provide the most robust solution. Concealed threat targeting could involve formulating a fire control solution against targets behind walls or abutments; the ability to engage unseen targets is clearly a decisive advantage for the squad. In each of the aforementioned cases, system calibration is critical to achieving the specified accuracy; calibration must account for internal system imperfections as well as changing environmental conditions.

Table G-2 also summarizes precision targeting missions. As in the two preceding sub-subsections, some sensors are dual-purpose. Additionally, precision targeting can maximize the capability of SA and force protection sensors, varying the sensor mode and collection strategy to calculate a fire-control solution. This approach is very common in radar, where system resources are modified to best meet the requirements of each mode—for example, short dwells and rapid antenna scans for search versus longer dwells and focused antenna beams for refined track and engagement.

Summary of Squad-Level Missions

Table G-2 summarizes all of the squad-level sensor missions (situational awareness, force protection, and precision targeting). The first column places technologies in one of those three mission areas. Some technologies support multiple mission areas.

SENSOR TECHNOLOGY GAPS

In this section, the committee assesses the gaps in squad-level sensor technology. Technology is assessed using the following key: green, mature; yellow, development required; and, red, serious technical hurdles remain. Where applicable, the relevant programs are mentioned.

Squad-level sensor technology development should carefully consider the issues identified in Table G-1. Sensor SWAP and deployment and TCPED strategy are of foremost concern. The fundamental issues are evident: How can sensor technology seamlessly provide the right information to the squad without disrupting cognition required to carry out critical elements of the mission? Sensor scaling, improved algorithms/techniques and computing, and autonomous platform capabilities are important in this regard.

To support materiel development, a rigorous systems engineering approach is also critical and should include:

- Training of the acquisition workforce,
- Development of enterprise-wide analysis tools, and
- Government-owned open system architecture.

The next three sections briefly describe sensor technology gaps from the squad perspective.

Situational Awareness Sensor Technology

Table G-3 provides a SA sensor technology gap assessment. While established sensor technology is available for a number of these missions, the constraints of squad-level deployment and operation is a primary concern.

TABLE G-3 Situational Awareness Sensor Technology Gap Assessment

Mission	Description	Relevant Sensor Technology	Technology Gap Assessment
SA	Dismount detection and engagement	Radar, SIGINT, FMV, IRST, WAMI	Deployment platform, scaling, TCPED, autonomy
SA	Vehicle detection and engagement	Radar, FMV, IRST, WAMI, acoustics, seismometer	Deployment platform, scaling, TCPED, autonomy
SA	Through-wall surveillance	Radar, SIGINT	Deployment, robustness, CONOPS
SA	Foliage obscured target surveillance	Radar, LADAR, SIGINT	Radar aperture size
SA	Spectrum surveillance	SIGINT	Obscuration, deployment, TCPED
SA	Navigation in GPS-denied environments	Radiofrequency sensor technology	Multi-transmitter deployment
SA, force protection	Concealed weapons detection	Millimeter-wave radar, metal detectors, magnetometers	SWAP
SA, force protection	Life signs monitoring	Millimeter-wave radar, acoustics, lasers	SWAP, deployment, CONOPS

Dismount and vehicle detection and discrimination capability has been a focus of recent Joint Improvised Explosive Device Defeat Organization (JIEDDO), Defense Advanced Research Projects Agency (DARPA), Army, and Air Force RDT&E investments. Examples of radar programs are: DARPA-JIEDDO's Vehicle and Dismount Exploitation Radar; the DARPA-U.S. Army

FOliage PENetration Reconnaissance, Surveillance, Tracking and Engagement Radar (FORESTER); the DARPA-U.S. Army Affordable Adaptive Conformal ESA Radar; and the All-Terrain Radar for Tactical Exploitation of MTI and Imaging Surveillance (ARTEMIS) of the U.S. Army's Communication-Electronics Research Development and Engineering Center. EO systems include the U.S. Air Force Angel Fire wide-area persistent FMV system; the US Army Constant Hawk ISR payload; the U.S. Air Force multitelescope, Gorgon Stare WAMI system for the Reaper; and the DARPA-sponsored Autonomous Real-Time Ground Ubiquitous Surveillance Imaging System for use on the YMQ-18A (Boeing A-160 Hummingbird). Each of these sensor payloads is currently suitable for larger, Class IV UASs—like the Reaper, YMQ-18A, or Global Hawk—or for light surveillance aircraft, but concerns over best TCPED strategies remain because enormous quantities of data are generated by each of the various sensors. Moreover, these payloads require continued improvement to operate in diverse threat environments. The assessment of yellow in Table G-3 was arrived at as a result of substantial concern over a suitable deployment strategy for the squad. A sensor package scaled for a lower tier UAV and close-range operation might be a good idea. TCPED—especially processing/exploitation and dissemination aspects—and autonomous and clandestine operation remain areas for further evaluation and development.

Through-the-wall surveillance technology has been another area of focus for DoD investment. Key efforts include the DARPA RadarScope, DARPA's Visibuilding program, the U.S. Army Sense-Through-The-Wall program, and the U.S. Navy Transparent Urban Structures (TUS) effort. Visibuilding and TUS both had reach-goals that included mapping the interior of specific buildings of interest to identify hallways, stairwells, hidden rooms, weapons caches, etc. The RadarScope is a weapon-mounted radar used to detect motion and heartbeats behind a wall. The U.S. Army Sense-Through-The-Wall blended features of Visibuilding, TUS, and the RadarScope. With the exception of the RadarScope, the sensor CONOPS and deployment of through-the-wall systems remain a concern. One possible strategy is to deploy the sensor on a tripod; UAS and tractor-trailer deployments have also been considered. In the deployment, operator motion—leading to false detections and obscuring potential targets—is a critical concern. This technology is given a yellow rating since further development, scaling, and CONOPs best meeting the squad's needs are needed.

Obscured target detection has been the focus of a number of developmental efforts, including the aforementioned FORESTER and ARTEMIS programs, the U.S. Air Force Tanks Under Trees effort, and the DARPA-U.S. Army Jigsaw program, to name a few. Jigsaw is a three-dimensional ladar that maps beneath the foliage by moving through a large angle and poking through holes in the tree canopy. Additionally, ground-penetrating radar systems are commercially available and regularly used by the electrical utilities industry. This technology receives a red assessment, since radar is the preferred and most robust technology that nonetheless must operate at low frequency, generally in the UHF (450 MHz and below). These low operating frequencies require physically large

(20 feet or more) antenna systems for effective operation. LADAR, such as Jigsaw, can be more compact; such LADARs, however, are limited by the characteristics of the obscuration: As the foliage increases in density, Jigsaw performance degrades. The use of several smaller, electrically coherent sensors on low-tier, autonomous UASs may be an option to overcome the challenges of obscured target detection at the squad level.

Spectrum surveillance—tactical SIGINT—in support of the squad must provide information with low latency and operate effectively in diverse and spectrally congested environments. Urban and mountainous terrains result in signal multipath and signal obscuration. Spectral congestion is a result of the significant demand for spectral allocation; spectral management techniques include architecting wireless cells with disjoint frequency allocations that repeat after a specified number of cells. Airborne collectors "see" the many emitters on Earth's surface, averting line-of-sight issues but increasing the co-channel interference problem. Multichannel processing and near-vertical incidence collection geometry are mitigating strategies. This technology area receives a yellow assessment because the squad's specific needs—ease of deployment, operation in complex environments, autonomous platform operation, and advanced TCPED—are not readily addressed by current technology.

Navigation in GPS-denied environments has been the target recently of RDT&E investment. Specific programs have considered navigation in caves and below ship decks. The key challenge is the transmitter deployment. GPS is easily jammed owing to low signal strength and simple receiver design, and so a separate radio navigation satellite system is unlikely to be useful. A better strategy from the squad's perspective is to deploy transmit signal sources on several (generally four or more) low tier UASs, such as the ScanEagle or to set up a regional network using larger UAS platforms; squad members then could rely on lightweight navigation receivers based on modifications to commercial designs. Alternative approaches, such as active ranging, require a communications link back to the squad; at the same time less desirable navigation communication links could present blue force tracking information directly to the squad. This technology receives a red assessment since investment would be required to develop and implement an appropriate solution.

Concealed weapons detection technology is currently available. The National Institute of Justice, for example, has invested in handheld mmw technology to image weapons hidden under clothing. Airport security screening includes mmw scanners to image hidden objects. The challenge from the squad's perspective is to develop and deploy a low-SWAP, mobile capability with a CONOP useful to the squad. Handheld devices are plagued by operator motion, and tripod mounted devices are unfortunately fixed and still require calibration. Packing mmw technology in a useful form for squad operation remains an open issue. For this reason, this technology area is given a yellow assessment.

Life signs monitoring technology does not appear to be currently available. DARPA has made some investments in this area and requested proposals from a number of potential sources. Likely sensor technologies include mmw radar and laser. Both solutions would probably require a tripod mount to avoid operator-sensor interaction. Advantages of the radar solution include its ability to penetrate clothing and perhaps armor. The laser can detect small, repetitive motion consistent with respiration. While a technological solution seems viable, given the absence of a specific program or deployed product, this mission area receives a yellow assessment.

Gaps in Force Protection Sensor Technology

Table G-4 assesses the gaps in force protection sensors. As in the preceding section, squad-level constraints—specifically, SWAP, mobility, CONOPS, and ease of deployment—dictate a more pessimistic assessment of the currently available technology.

TABLE G-4 Force Protection Sensor Technology Gap Assessment

Mission	Description	Relevant Sensor Technology	Technology Gap Assessment
SA, Force protection	Concealed weapons detection	Millimeter-wave radar, metal detector, magnetometer	SWAP
SA, Force protection	Life signs monitoring	Millimeter-wave radar, acoustics	SWAP, deployment, CONOPS
Force protection	Perimeter surveillance	Radar, acoustics, SIGINT	Solutions currently available, reduced SWAP desirable
Force protection, precision targeting	CRAM	Radar, IRST, acoustics	SWAP, mobility
Force protection	CIED	Radar, HSI/MSI, SIGINT	Challenging target signature, persistence
Force protection, precision targeting	Counter-shot/sniper	Acoustics, IR	Multipath, calibration, SWAP

Mission	Description	Relevant Sensor Technology	Technology Gap Assessment
Force protection	Mine detection	Ground-penetrating radar, HSI/MSI, magnetometer, metal detector	Commercially available solutions, vehicle mounted
Force protection	CBRN agent detection	CBRN-tailored sensors, remote sensing techniques (radar, EO/IR, HSI/MSI)	Customized sensor development for varying threat characteristics

The first two lines in Table G-4 were discussed in prior sections. Perimeter surveillance technology is well developed and has been used in the field for several decades. From the squad-level perspective, miniaturization, power reduction, and automation efforts would prove most beneficial.

The U.S. Army has invested substantially in CRAM technology. Systems like Firefinder and Enhanced Firefinder are sophisticated, weapons-locating radar systems. The Omni-Directional Weapons Location radar is a new capability being developed by PEO IEWS/PM Radars. The Firefinder and the Omni-Directional Weapons Location radars are fairly large systems with significant prime power requirements. For this reason, they have limited applicability at the squad level. The Lightweight Counter Mortar Radar weighs approximately 90 lb and is packed in two sections; its ruggedized design enables it to accompany paratroopers on airborne assaults. Yet, 90 lb is still substantial load for squad members. A smaller, lighter, shorter-range version of the Lightweight Counter Mortar Radar would better support the squad. This new, lightweight system should provide 360 degree coverage and accept battery power. Moreover, the system should be highly transportable, with minimal set-up time. Taking advantage of novel materials and electronics should be an imperative in this new CRAM system design. The U.S. Army has also invested in active protection systems (APSs) to protect ground vehicles and rotorcraft. Cost and less-than-hoped-for cooperation of our allies have proven major challenges in deploying APS's, along with concerns over anti-radiation seekers. Identifying a way to integrate vehicle and dismount detection missions with CRAM is a meaningful objective; APS will likely have to be a unique sensor package tied to a kinetic kill mechanism. Integrated Force Protection Capability (IFPC) is a new program of record focused on multisensor integration of CRAM products; networking capabilities developed under this program may find applicability to squad-level protection.

CIED is a three-pronged approach. Two of the prongs are direct: Find the IED and jam its triggering mechanism. The third prong is indirect and centers on finding the network that supplies and emplaces IEDs. Squad-level missions benefit from the direct CIED approaches. ISR technology can be used to find potential IED emplacements: This information can be provided directly to the squad. Yet, providing technology to the squad that allows remote status monitoring of culverts and other structures where IEDs can easily be emplaced is more useful. For example, providing the squad with radio frequency or optical readers to scan antitamper mechanisms along a chosen route is a direct, valuable, and low-SWAP solution, especially if integrated with other systems. IED electronic warfare technology, such as the Joint Counter-Radio-Controlled Improvised Explosive Device Electronic Warfare system, is effective for vehicle-borne squad missions and should continually be improved.

The DARPA Boomerang system is a counter-fire, small arms locator. It was originally developed for use on vehicles. Specifically, it was found that vehicle noise made it difficult for blue forces to identify the location of hostile, small arms fire. Boomerang, a multimicrophone system, would provide a general location of incoming fire. Multipath signal degradation, especially in urban and mountainous terrain, is a fundamental, limiting factor. The extension of Boomerang to the dismounted squad was a planned activity under the Land Warrior system; this integration has not yet been accomplished, perhaps in part due to cancellation of Land Warrior. Providing enhanced small arms locating systems to the squad should be an objective. The lack of such a capability and the degradation of sensor performance in urban and mountainous terrain lead to a yellow assessment for this sensor mission area.

A number of commercially available ground-penetrating radar systems are available. Mine detection is complicated by mine composition and soil attributes. Naturally, detecting a metal mine in dry sand is easier than detecting a composite mine in wet clay. Generally, ground-penetrating radars are vehicle mounted and usually placed in proximity to potential mine locations. Other than vehicle-borne ground-penetrating radar, it is hard to imagine a dismounted Soldier mine detection capability, except for the very dangerous approach that employs metal detectors. This technology is mature, but the hatched green assessment in Table G-4 indicates that it may not be possible to further adapt mine detection capability to the squad.

Technology is currently available to respond to an array of chemical, biological, radioactive, and nuclear agents. For example, it is possible to build carbon nanotube switches that are sensitive to ammonium nitrate or a number of other chemicals; once the switch is thrown, a signature characteristic of the deployed device—such as resonant frequency—is detectable via remote sensing by means of, for example, radiofrequency or optical probing. The development of low-cost CBRN detectors that are reliably and easily probed by a squad during execution of its mission is an invaluable force protection capability. Some relevant technology has been developed and demonstrated in government and university laboratories, and further system development is warranted. A responsive approach that is able to deploy new sensors as the threat evolves is essential, hence the yellow assessment.

Gaps in Precision Targeting Sensor Technology

Table G-5 provides a gap assessment of precision targeting sensor technology. The first two rows were discussed previously in the preceding section.

TABLE G-5 Precision Targeting Gap Assessment

Mission	Description	Relevant Sensor Technology	Technology Gap Assessment
Force protection, precision targeting	CRAM	Radar, IRST, acoustics	SWAP
Force protection, precision targeting	Counter-shot/sniper	Acoustics, IRST	SWAP, deployment, CONOPS
Precision targeting	Vehicle engagement	IR, radar, optical sights	Technology available
Precision targeting	Concealed threat targeting	EO/IR, radar imaging	Specialized sensors coupled with new weapons needed

Vehicle engagement at range is presently supported by radiofrequency, EO, and IR seeker technology. Optical sights can be used to support long-range operation. Ancillary sensors to measure environmental conditions may be necessary. Overall, sensor engagement technology is mature and enjoys rich collaborative efforts across the Services and with coalition partners. One area for consideration at the squad level is correctable projectiles. The DARPA Self-Correcting Projectile for Infantry Operation program integrated sensor technology and piezo-based actuators into a 44-mm projectile to compensate for dispersion due to muzzle velocity variation. Perhaps coupling like technology with offboard sensor information to engage targets at long distance with modest caliber projectiles would vastly boost dismounted squad lethality.

In this vein, moving target indication, discrimination, and tracking capabilities or fixed-target imaging systems are critical. Mounting corresponding capability on a UAS provides an approach for peering behind abutments and buildings and the like. The U.S. Air Force has demonstrated synthetic aperture radar target geolocation and hand-off to GPS-guided munitions for precision targeting; a famous video of a smart munition entering an elevator shaft during the first gulf war points to the maturity of this radar and navigation technology. Newer capabilities, such as the DARPA Synthetic Aperture Ladar for Tactical Imaging (SALTI) system may prove more useful on small UASs supporting the squad; the SALTI system's goals include high resolution, coherent, optical imaging with three-dimensional views. The Global Hawk UAS served as the target platform for the SALTI payload. Scaling SALTI to smaller UASs may be possible. In Table G-5, this technology receives a red assessment, since no autonomous capability is currently available or envisioned. The preceding line on vehicle engagement is green hatched, since technology is available, but it has not been adapted to squad-level activities.

Appendix H

Prospective Robotics Technologies

This appendix describes robotics technologies and approaches being developed to contend with the issue of load carried by dismounted Soldiers and to extend operational capabilities of the tactical small unit (TSU). The state-of-the-art technology readiness level (TRL) is provided for each of the advanced system examples.

WHEELED ROBOTIC SOLUTIONS

Wheeled robotic solutions are robotic systems that mimic the off-loading function served by mules and donkeys around the time of World War I.

Example 1

The Future Combat Systems Multifunction Utility/Logistics and Equipment vehicle is a 3.5 ton, six-wheel vehicle built by Lockheed Martin; it can be reconfigured from a logistics carrier to an automated weapons platform. It is teleoperated with limited autonomy, such as leader/follower or following electronic breadcrumbs. It is capable of operation in difficult terrain. The development was at TRL state-of-the-art 6 when the program was canceled in 2010 due to disappointing field trials and high projected cost.

Example 2

The Israel Aerospace Industries' REX is similar in size and function to the Future Combat Systems Multifunction Utility/Logistics and Equipment. A small version of the REX is a four-wheel all-terrain vehicle weighing approximately 350 lbs. and carrying 500-lb payload. The unit can be configured for a number of missions and has leader/follower capability. It follows Soldiers using electronic breadcrumbs, can understand hand signals and avoid obstacles, and it exhibits multisensory position determination. The small REX has a sixty-mile range on a single tank of fuel and is currently being actively marketed for applications at the

Soldier level. Its development potential is TRL 7 in the mid term and TRL 9 in the far term with full autonomy.

BIPEDAL ROBOTIC SOLUTIONS

Bipedal robotic solutions include exoskeletons and anthropomorphic robotics for load-bearing, enhanced-lifting, and increased-endurance, and a stand-alone bipedal robot.

Example 1

With funding from the Defense Advanced Research Projects Agency (DARPA), both Lockheed Martin, and Raytheon developed exoskeleton systems that use imbedded computers and sensors to determine what the wearer wants to do and moves the powered elements accordingly. Units are battery powered and transfer load and weight of system to ground through lightweight, high-strength lower limb elements. Both have been shown to augment many human functions such as walking/running, squats, crawling, upper body lifting.

Both are undergoing further development to improve energy efficiency and evaluate their effect on human performance and user acceptance. Development potential at present is TRL 5, advancing to TRL 6 in the mid term and TRL 9 in the far term with other planned robotic functions including navigation, mission planning, and heavy lifting in field. First implementation is likely to be in rear/depot areas where moving of heavy objects is prevalent and energy supply less critical.

Example 2

The Hybrid Assistive Limb® (HAL) exoskeleton was developed by the Japanese company Cyberdyne. HAL is a cyborg-type robotic exoskeleton that has been developed mostly for the medical market. It is controlled by reading bioelectric signals on the human limbs that tell the muscles what to do and when to do it. This signal is fed to a computer that autonomously directs the exoskeleton to carry out the desired function. It is promoted as providing a more than two-fold enhancement of the human potential for activity. It is battery powered and specified to operate for more than 2 hours on a single charge.

Status: Currently available commercially in limited quantities. Available in lower limb and full-body cyborg like configuration. TRL is 6-7 overall.

Development Potential: Commercial units not designed for military applications. Development potential excellent.

Example 3

Boston Dynamics and Honda are both experimenting with full anthropomorphic bipedal robots. Figure H-1 shows the Boston Dynamics robot in its current stage of development without any external shell.

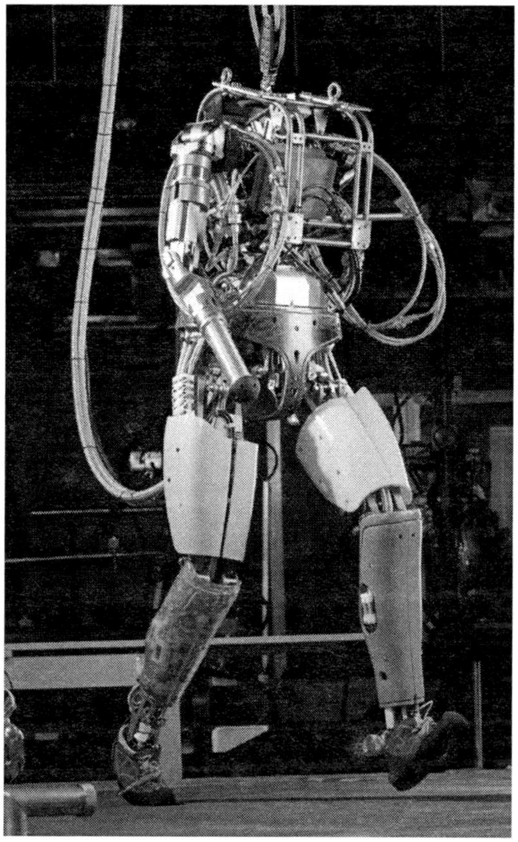

FIGURE H-1 Protection Ensemble Test Mannequin (PETMAN) robot without external shell. SOURCE: Re-printed courtesy of Boston Dynamics.

It is designed to function with most of the mobility of humans. The robot, called Protection Ensemble Test Mannequin (PETMAN), is being developed with Army support for testing clothing in a biological and chemical warfare environment. The robot is capable of walking, crawling, doing calisthenics, and most general human mobility functions. It is hydraulically actuated and has shock absorber elements in its legs. Control algorithms for the PETMAN are derived from the Boston Dynamics series of quadruped robots described below.

Status: TRL 5 proof of principle laboratory robots are under development and testing.

Developmental Potential: Excellent. Mid term TRL 6 with expanded applications. Fully functional anthropomorphic autonomous robots are anticipated for the far term.

Example 4

The Japanese robot Advanced Step in Innovative Mobility (ASIMO), is a bipedal anthropomorphic robot by Honda. ASIMO is in a sense a test-bed machine for developing and improving the autonomous behavior of an anthropomorphic robot. It looks somewhat like a human and performs like an intelligent entity. In the latest version, ASIMO is claimed to have advanced from a rule-based automatic machine to an autonomous "machine" able to function in a social environment and make decisions in the context of its surroundings.

According to the Honda Web site,

> The following three factors were identified as necessary for a robot to perform as an autonomous machine, and the technologies required to realize these capabilities were developed: 1) high-level postural balancing capability which enables the robot to maintain its posture by putting out its leg in an instant, (2) external recognition capability which enables the robot to integrate information, such as movements of people around it, from multiple sensors and estimate the changes that are taking place, and (3) the capability to generate autonomous behavior which enables the robot to make predictions from gathered information and autonomously determine the next behavior without being controlled by an operator.

ASIMO is 1.3 meters tall, weighs 48 kg, and exhibits 57 degrees of freedom. It has demonstrated its ability to do the following:

- Establish a "world view" and perform within it through multisensory inputs and artificial intelligence programming.
- Ascribe intent to the movement and people and predict future configurations.
- Recognize multiple faces and voices simultaneously and change behavior within context.
- Predict trajectories and plan intercept path with subsequent voice engagement.
- Walk, run (9 km/hr), run backward, hop on one or two legs, and traverse uneven terrain.
- Tactile sensing for preforming delicate tasks.

Status: Highly developed and fully functional test bed for autonomous behavior in robotic systems. TRL of 6-7.

Development Potential: Excellent. In the far term, robotic systems based on ASIMO technology could produce humanoid robots for military applications.

QUADRAPED ROBOTIC SOLUTIONS

Quadraped robotic solutions are the robotic equivalent of donkeys or mules as pack animals. When fully developed quadruped robotic systems should have more mobility than wheeled or tracked vehicles. While several "legged systems" are being developed, the two described below are the most advanced. Examples include:

Example 1

BigDog is an advanced quadraped robot built by Boston Dynamics under contract to DARPA (Figure H-2). It has demonstrated the ability to traverse difficult terrain and can walk, run, and climb steep slopes while carrying respectable loads. BigDog's power source is a gasoline- powered small engine that drives a hydraulic system to actuate all of its mobility functions. The legs of the robot are compliant and can recover some of the energy normally expended in the shock and bending associated with walking/running. BigDog is about the size of a small donkey, about 3 ft long, 2.5 ft tall, and weighing 240 lbs. BigDog has an onboard computer that contains a sophisticated control system to actuate all mobility functions, health monitoring, and advanced sensing and position location functions. In its current embodiments, it is teleoperated.

FIGURE H-2 BigDog. SOURCE: Re-printed courtesy of Boston Dynamics.

Boston Dynamics claims that BigDog has demonstrated the following:

- Running at 4 mph,
- Climbing slopes of up to 35 degrees,
- Traversing rubble-strewn terrain,

- Mobility in rain, mud and snow, and
- Ability to carry a 340-lb load.

Current Status: Currently TRL 6 in this embodiment as a technology demonstrator.

Development Status: Focus has shifted to larger sizes in the next example described.

Example 2

The Legged Squad Support System (LS3) is a scaled-up version of BigDog with bigger payloads and more capability. See Figure H-3.

FIGURE H-3 LS3, Legged Squad Support System. SOURCE: Re-printed courtesy of Boston Dynamics.

LS3 is in the early stages of development with DARPA funding and is advertised as far more capable and having greater range, reduced acoustic signature, and the ability to operate in a follower mode with troops.

Current Status: TRL 5-6. First items are being evaluated by DARPA.

AUTONOMOUS SYSTEMS

Teleoperated robotic systems are in widespread use by Soldiers, but these require constant human control to operate, which can degrade the performance of small units in combat. Beyond teleoperation, numerous experimental robotic systems have been built to demonstrate specific aspects of advanced robotic functions such as autonomous navigation over extended distances, a rudimentary world view and operation within that view, anthropomorphic bipedal mobility, quadruped mobility, merging of human and robot through a powered exoskeleton, and human-robot interaction. All are promising but none have been developed to the point of practicality the context of military applications.

The cutting edge for robotics is a fully autonomous system and all it implies: Perception, world view, human-robotic interaction, and the rest are equally important for robotics to mature to its full potential. There have been many instances where robotics seem to exhibit autonomy in scripted and highly supervised scenarios. To date, however, there has not been an instance of a *learning* robotic entity, perhaps confined to a military base initially, that is allowed to roam freely, mingling with humans and human-driven machines, while performing assigned duties, as would be required to demonstrate full autonomy. The most impressive demonstrations of reasonable autonomy from the perspective of the military have been the DARPA challenges where robotic systems traversed long distances over varied terrain given only a starting point and an end point.[1]

Building on the DARPA grand challenges success, Google has started a major initiative in advanced robotics and to date has logged over 140,000 miles in a robotic vehicle with impressive results. Google usually allows autonomous operations on roads that have been traversed by a research team to develop data from which to generate a "world view" of the intended route. In addition, Google uses cloud computing to give the vehicle the advantage of a much larger computational capability than could be housed in the vehicle itself. Even so, the results are impressive and represent a step toward autonomy on a par with the far more modest technology displayed by ASIMO described above.

Large Unmanned Air Vehicles

Large unmanned aerial vehicles (UAVs) are used to support Soldiers and squads by providing for reconnaissance and intelligence and, in the future, for delivery of lethal agents. For maximum utility, the overhead asset could be tasked by a Soldier or squad leader to observe a forward area and deliver data and/or photo images directly and in real time. If a critical threat exists, the UAV could deliver a weapon to take out the threat. All of this capability exists today with the exception of the communications link and doctrine that would enable the Soldier to access and task the UAV.

---

[1]Additional information is available online at www.darpagrandchallenge.com.

A wide range of large UAVs in various stages of development and having loiter times in excess of 7 days are applicable to this capability (DoD, 2009). The most notorious of those in current service is the Predator, which was introduced in 1997 by the Air Force. Since that time, it has flown thousands of missions, including surveillance and weapons delivery missions, and is capable of flying at over 25,000 ft for over 20 hr. At such altitudes, these systems are less susceptible to the weather and environmental factors that prevail closer to the battlefields. In the far term, high resolution Synthetic Aperture Radar could give the high-altitude UAV all-weather capability but with reduced resolution and fidelity of data. The systems are teleoperated from remote locations by trained pilots. Since communications and data transfer exist for these systems, the only remaining obstacles to tasking at the squad level are to put in place the necessary communications links and to make the UAV available for a specific mission set.

Current State of the Art: TRL 9 currently deployed in theater and by the U.S. Customs and Border Patrol

Small UAV and UGV Robots

Robots practical for use at the squad and individual Soldier levels are of two kinds: those that have to be carried by the Soldier and those that can keep up using their own motive power. At the Soldier level, robotic weight and size restrictions are tied to mission. For surveillance and reconnaissance several lightweight UAVs and UGVs (unmanned ground vehicles) that could be applied at the squad level are currently in the inventory or being evaluated in the field:

- *UAV.* Gas Micro Air Vehicle (GMAV) ducted fan, 16 lb, 3-lb payload, liquid fuel.
- *UAV.* Wasp III winged, 1 lb EO/IR sensor suite, battery.
- *UAV.* RQ-11 Raven winged, 4.2 lb, 11.2-oz payload, battery.
- *UGV.* Packbot, tracked, ~30 lb, 4-lb multifunctional payload, battery
- *UGV.* MARCbot, wheeled, ~25-lb multifunctional payload, battery
- *UGV.* Throwbot, wheeled, 12 oz, battery.
- *UGV.*Toughbot, wheeled, 2.1 lb, battery.

The GMAV ducted fan system is able not only to hover over a target but also to perch on top of buildings or other structures to provide persistent surveillance. The Throwbot and Toughbot UGVs are designed to be tossed into an area such as a building or enclosure to provide video surveillance of the area.

Small UAV and UGV systems can provide the Soldier with the ability to look beyond the next hill or into the next block of an urban environment, identify enemy positions, explore buildings and caves, and locate and examine potential improvised explosive devices. All of these tasks are dangerous and account for numerous casualties if done by Soldiers.

REFERENCE

DoD (Department of Defense). 2009. FY2009–2034 Unmanned Systems Integrated Roadmap. Washington, D.C.: Department of Defense.

Appendix I

Energy Technologies and Applications for the Soldier

With the exception of harvested solar and wind energy, almost all of the Soldier's energy supply comes from the energy stored in chemical bonds. That energy is accessed by the Soldier in the form of food, explosives/propellants, and electricity from batteries, fuel cells, and fueled systems, such as internal and external combustion engine/generator combinations. Figure I-1 shows common energy sources in terms of both specific energy and energy density.

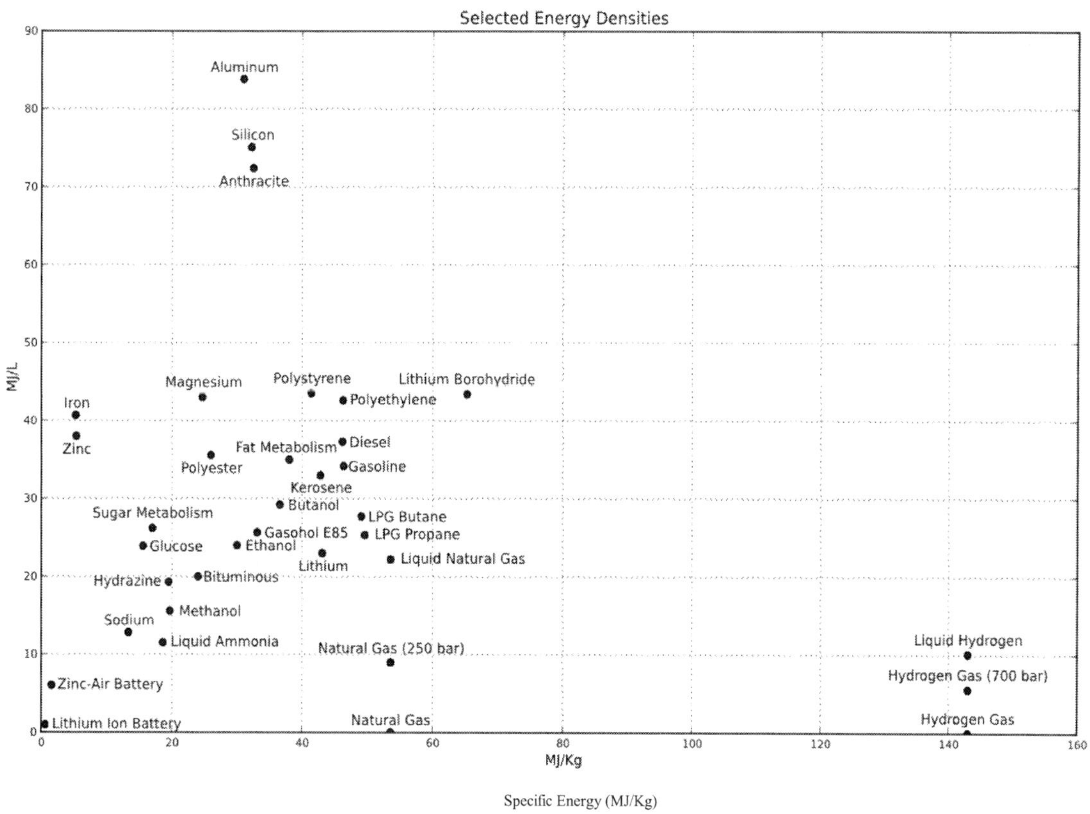

FIGURE I-1. Selected energy densities. SOURCE: http://en.wikipedia.org/wiki/File:Energydensity.svg.

While the energy content of these sources may be quite high, the amount of that energy that can be efficiently extracted is only a fraction of what is theoretically possible for most applications. Table I-1 is a compilation of the most relevant technologies and current states of the art.

TABLE I-1 State of the Art for Technologies Most Relevant to the Dismounted Soldier

Battery Chemistry	Specific Energy Theoretical	(Whr/kg) Battery	Comments/State of the Art
Primary batteries			
$LiSO_2$	1,175	190	Inventory item BA5590
$LiMnO_2$	1,000	220	Inventory item BA5390
$Li(CF)_x$	2,180	370 - 600	Inventory: low specific power version; in half BA5590 size at TRL 5-6
Secondary batteries			
Lithium polymer	750	130-200	Commercial product. Widely used in cell phones.
Lithium ion	750	108	Inventory BB5590
$LiCoO_2$		140	Inventory LI-145
		158	Inventory LI-80
Zinc-air	1,370	280-300	Inventory BA-8140, BA 8180
Lithium-air	5,210 (including oxygen)	800 cell demo; 400 half battery	TRL 4-5 experimental. Rechargeable possible at TRL 4

The Communications-Electronics Research, Development, and Engineering Center (CERDEC), the Natick Soldier Research, Development, and Engineering Center and the Army Research Laboratory have research and development programs to overcome current shortcomings of energy sources for the Soldier.

BATTERY TECHNOLOGIES

The trend in batteries, partially driven by commercial applications and the potential for high specific energy, is with a few exceptions, toward lithium chemistry (both disposable and rechargeable batteries). The two notable are exceptions (1) the zinc-air system, which is a primary disposable battery, is an "air breather" (as are fueled systems), and is fielded in a variety of versions and (2) thc lithium air battery, which is currently in the early stages of development and can function as a primary or secondary battery.

Prototypic systems that will undergo field evaluation in the near term include the following:

- A high-power, high-specific-energy conformal centralized primary lithium battery system.
- Primary lithium batteries based on $LiMnO_2$.
- Primary batteries based on $Li(CF)_x$ (half size and low specific power).
- The Net Warrior rechargeable Li 145 lithium ion battery.

In the near term, the specific energy and energy density of these batteries are still far removed from what is theoretically possible. The wide disparity between what is theoretically possible for batteries based on lithium chemistry and what is practical shows that there is much room for improvement. The technical challenges to packing more energy into smaller size are always further complicated by practical considerations such as safety, life, charge/discharge rates, and cost.

Ideally, one would like the rechargeable batteries to be as energetic as the primary single-use batteries. Rechargeable batteries are about 75 percent as energetic as primary batteries. Battery technologies such as lithium polymer, lithium ion polymer, and lithium ion batteries can be improved substantially through the development of new electrolytes. In addition, the engineering of more active electrode materials and innovative packaging can result in major improvements in stored energy without sacrificing the other characteristics demanded by users. For both primary and secondary batteries, cell-level tests indicate that batteries can be improved by a factor of 2 or more to approach 300 Whr/kg in the near to mid term (5-10 year time frame), thereby reducing weight and volume.

For the far term (10-20 years), the development of both primary and secondary versions of Li-air energy sources will be revolutionary and offer performance on a par with that of fueled systems. The basic chemistry of Li-air has been demonstrated in both battery formats (see, for example, Laoire et al., 2011; Zheng et al., 2008). A survey of the technical literature identifies the following research and engineering issues that must be resolved to reduce the technology to practical embodiments:

- Detailed quantitative understanding of the electrochemical kinetics of charge/discharge cycling since these determine chemical reversibility and the level of coulombic efficiency in cycling.
- Development of appropriate electrolytes and cathode structures that can function in the presence of O_2.
- Development of electrocatalysts.
- Development of high-porosity air cathodes that maximize transport of reactants to the catalyst surface.
- Development of a suitable lithium metal or lithium composite electrode that minimizes dendritic growth and environmental contamination.
- Development of air separator membranes that allow oxygen to pass while excluding environmental contaminants and water.

As a measure of progress, CERDEC-sponsored programs in lithium-air batteries have achieved impressive results in a nonrechargeable format. Energy densities of 800 Whr/kg were achieved for individual cells, and in a "first packaged" technology readiness level (TRL) 4 demonstration, a multicell lithium air battery in a half -90 configuration achieved 450 Whr/kg (Laoire et al., 2011). Forty charge/discharge cycles with 100 percent coulombic efficiency have been reported.

FUEL CELL TECHNOLOGIES

Fuel cells have been researched for about as long as batteries. Their primary attributes are high conversion efficiency, low acoustic and thermal signatures (some embodiments), and favorable scaling over a wide range of sizes. They lag behind batteries in practical embodiments, however, primarily because they require fuels that are not currently in the military logistics chain. In its Appendix D, the NRC study *Meeting the Energy Needs of Future Warriors*, provides a comprehensive comparison of the basic technologies and performance parameters for relevant fuel cell technologies (NRC, 2004). Like zinc-air and lithium-air batteries, fuel cells are "air breathers," which places restrictions on immersion and necessitates filtering in extremely dusty environments.

Direct Methanol Fuel Cells

Direct methanol fuel cells (DMFCs) are proton-exchange fuel cells in which methanol, CH_3OH is used directly as the fuel. Their main advantages are as follows:

- Methanol is a commonly available fuel.
- Methanol is stable over a wide range of environmental conditions.
- It is easy to transport.
- The nonpolluting waste is mostly water vapor with carbon dioxide.

The main disadvantages are these:

- Methanol has a low specific energy compared to aviation fuel and gasoline.
- It has a low conversion efficiency (thermodynamic efficiency ~40 percent).
- Complex balance of plant.
- Its introduction into the military inventory would require a new and separate fuel infrastructure.

Reformed Methanol Fuel Cell

Reformed methanol fuel cells (RMFCs), are proton-exchange fuel cells, where the methanol is reformed into a hydrogen gas stream before entering the fuel cell. The advantages of RMFC over DMFC systems are as follows:

- Higher efficiency.
- Better low-temperature performance.
- Simplified stack management (no water).
- Higher specific power.

Their main disadvantages are these:

- Waste heat must be managed.
- Methanol reformers are complex.

Solid Oxide Fuel Cells

Solid oxide fuel cells (SOFCs) operate at temperatures high enough to function on a wide range of hydrocarbon fuels without external reforming. This is due to the high internal temperatures that allow fuel oxidation at the anode and largely negate the need for catalyst. The electrolyte in a solid oxide fuel cell is a hard, nonporous ceramic compound that allows versatility in stack construction. SOFC advantages are these:

- High-temperature operation allows internal reforming.
- They are highly resistant to poisoning by sulfur and carbon monoxide.
- Versatile geometry.
- High efficiency.
- Multifuel capable.

Its basic disadvantages are these:

- The high temperature calls for special materials.
- Slow start-up times.
- High temperatures require special thermal management techniques.
- Reliability and robustness are issues.

Comparison of Fuel-Cell Types

Table I-2 lists the fuel-cell-driven technologies that have been developed for military applications. The last column in the table provides current TRL and status.

TABLE I-2. Fuel Cell State of the Art

Type and Fuel Dry Weight	Power level (Watts)	Efficiency %	Dry System Specific power (W/kg)	Start up time (min)	Mass for a 72-hr mission (kg)	Comments
DMFC Methanol Water/methanol mix at T>40C 1.18	20-25	22.4	21	10	2.6	TRL 7 Soldier power source (field evaluation)
RDMFC Methanol 1.6 kg	50		31	30	18	TRL 6-7 Hybrid or stand-alone energy source (field evaluation)
SOFC Propane fuel 2.6 kg	50	50	19.	30	6.2	TRL 6-7 Hybrid or stand-alone power source (field evaluation)
DMFC Methanol/water mix 15 kg	300	22.4	20	10	Ops specific – would not be required to run continuously	TRL 6-7 Squad battery charger and stand-alone power supply
RMFC Methanol 16.3 kg	300		18	30	Ops specific – would not be required to run continuously	TRL 6-7 Squad battery charger and stand-alone power supply
SOFC Propane 16 kg	300	50	19	15	Ops specific – would not be required to run continuously	TRL 6-7 Squad battery charger and stand alone power supply

In the mid-term (5-10 years), the basic technologies associated with the fuel-cell types in Table I-2 have already undergone extensive development and aside from reducing the cost of manufacturing, the biggest single gain that can be expected is to develop a reliable, sulfur-tolerant jet propellant (JP) fuel reforming capability that could be integrated into these fuel cells. Note that both liquid propane and JP fuels have approximately the same specific energy but that liquid propane has about half the energy density in Whr/liter that JP has. Both have roughly 2.25 times the specific energy of methanol, a difference that increases when it is necessary to use a methanol/water mixture for operation at higher temperatures.

As with any of the fueled systems, there are two issues associated with warm-up time and with operation of the small fuel cells as part of a hybrid system that the Soldier can carry: immersion in water and contaminant ingestion. DMFC and RMFC are currently being built for limited deployment in Afghanistan and in other active military operations. All three of the main types discussed above are at a TRL of 6-7 and are in various stages of testing. As mentioned, their widespread introduction into the inventory will also require the introduction of a new "logistics fuel," which does not seem to be practical at this time.

In the far term (10-20 years), the technology using hydrocarbon fuels will continue to mature and to achieve marginal increases in performance. However, given that hydrogen has a specific energy of about 40,000 Whr/kg, roughly 3.25 times that of JP and liquid propane gas, major technological advances in a highly competitive civilian hydrogen economy could drive increases in system-specific energy greater than JP fuel systems. Note that the energy density in Whr/liter is extremely poor, forcing trade-off between mass and volume for specific hydrogen storage technologies.

COMBUSTION SYSTEMS

In two earlier studies, *Energy-Efficient Technologies for the Dismounted Soldier*, (NRC,1997) and *Meeting the Energy Needs of Future Warriors*, (NRC, 2004), fueled electrical energy sources were considered in the size range relevant to the dismounted Soldier. In general, the conclusions from those studies show that fueled sources have significant potential for providing a reliable source of energy for the Soldier. In both studies, conversion efficiencies in the 10-50 percent range (fuel heat value to usable energy) were possible, but the level of technical development for suitable fuels was at a low TRL.

The Army is rapidly moving to implement rechargeable battery technology as the Army standard as the specific energy of rechargeable batteries improves and the costs of primary batteries and of their delivery to theater falls. While delivery of a single battery to theater is approximately the same no matter what the type, rechargeable batteries will cost about 5 times as much to manufacture. Given that they can undergo the charge/discharge cycle 200 times or more—a primary battery is used only once—it is obvious that the basic cost and delivery cost savings are enormous. However, this scheme necessitates having battery chargers at several points, down to the level of vehicles and the individual Soldiers themselves.

The Army has begun development on a recharger technology in limited production using fuel-cell prototypes. Unfortunately, the fuel cell alternatives use hydrogen, propane, or methanol fuels, and the Army would much prefer not to have a new battlefield fuel in the inventory. As an alternative to fuel cells, CERDEC funded research on two types of systems that could use standard battlefield fuels—small external combustion Stirling energy systems and small internal combustion engine converter systems.

External Combustion—Stirling

External combustion engines such as steam engines and Stirling cycle engines have been in use since about 1800 but seem, except in a few embodiments, to have mostly been relegated to history, because internal combustion engines and electrical power from an ever-expanding grid were more efficient. Their primary advantage lies on the fact that the thermal process is steady state, which allows combustion optimization and energy recuperation. Further, steady-state combustion inherently has a lower acoustic signature than internal impulsive combustion. It is possible to operate two separate free-piston versions of Stirling engines such that all vibration is canceled, resulting in an extremely quiet system.

Early versions of Stirling engines employed exotic materials and had low specific power even though they were efficient converters of thermal energy to electricity. In recent years, however, advances in materials have led to the development of components with sufficient high-temperature properties that interest in Stirling technology has emerged as a viable energy converter for some applications. It is currently a viable candidate for deep space exploration[1] and shows promise for battlefield (NRC, 2004) and commercial applications such as co-generation.[2] Since the Stirling converter requires only a heat source, it is inherently multifuel-capable and has been demonstrated with a range of energy sources, from nuclear to heavy distillates. The Stirling engine can be made in a range of sizes with no loss of efficiency. For example, since about 1990, free-piston Stirling engines have been successfully demonstrated over power levels from 40 W to 25 kW. Stirling engine technology offers the potential for unique military systems that have extremely long life (>10 years) of continuous operation, unprecedented reliability in a military environment, and extremely simple and elegant mechanical functionality—the motor alternator comprises only one or two moving parts, neither of which is in contact with the other. Since 2005, there has been steady progress in taking the NASA investment in Stirling technology for deep space missions and adapting it for use as a battlefield energy source (CERDEC-funded Defense Advanced Research Project Agency (DARPA)

[1]Additional information available at http://www.grc.nasa.gov.

[2]Additional information available at http://www.stirling-tech.com/cogeneration.

Palm Power program). Proof-of-principle prototypes at a TRL of 5-6 were built and tested.

The Army has been funding the development of small Stirling converter technology for a number of years based on its desirable characteristics and the fact that it could be made man portable and multifuel capable. The Stirling engine ultimately promises to deliver:

- A basic motor alternator with >20,000 hrs of continuous life.
- Mean time between failure of balance of plant items >5,000 hrs.
- Replacement needed only in the case of extreme abuse or battle damage.
- Most spare parts would be black boxes such as controllers.
- Logistic fuel powered and can work on any heat source.
- High efficiency minimizes battlefield fuel requirement.
- May use waste heat in combined heat and power applications >80 percent efficiency.
- Battery charger makes rechargeable batteries a real possibility.
- Silent watch applications for long duration.
- Acoustic signature of less than 50 dBA at 7 m.
- Can operate in any orientation.
- Operates on the move.

Small Internal Combustion Engines

Soldiers in Iraq and Afghanistan purchased commercial Honda generators in the 1-3 kW range to augment their energy sources. These units use gasoline but can be modified easily to function with propane. They have a limited format for the electrical output: 12 V DC and 120 V AC. Noting that commercial Honda generators cost less than \$1.00/W, there have been numerous efforts in the past 10 years to modify commercially available engines, both two cycle and four cycle, to run on heavy distillates.

CERDEC's small internal combustion engine development efforts use standard gasoline fueled-engines. The first approach to converting small gasoline engines to heavy fuel use is to purchase a standard, hand pull-start commercial Honda generator and modify the carburetor to accept vaporized heavy distillate. There are three overriding problems with this approach: (1) the fuel, although vaporized, still consists of long-chain hydrocarbon molecules that are hard to ignite, (2) partial combustion leads to coke formation and gumming of the engine—wet stacking—and (3) the engine is difficult to start.

The alternative CERDEC approach is to develop a fuel "gassifier/cracker" that breaks up the large molecules in heavy distillate fuels at the expense of a few percent decrease in energy density. Such a fuel cracker was developed initially for Honda by Precision Combustion, Inc. (PCI). Basically, the output from the gassifier contains a high percentage of molecular hydrogen (H_2), CO, and CO_2, as well as molecules like C_3H_6, in various proportions. The focus of the PCI effort is to miniaturize the "gassifier/cracker" to suit a small engine rather than an engine like the Honda one, which produced many kilowatts of thermal power. This device is essentially the first stage of a conventional reformer and the basic physics is understood. The hydrogen in the gas stream should aid starting; the smaller molecules should limit coking; and the technology will build on previous successful efforts for Honda.

Combustion Systems State of the Art

The near-term state of the art for combustion technology developments is depicted in Table I-3. The last column lists the levels of development in terms of TRL with remarks on the applicability to TSU operations.

TABLE I-3 Near-term State of the Art for Relevant Combustion Technologies

Type and Fuel and Dry Weight	Power Level (Watts)	Efficiency (%)	Dry Systems specific Power W/kg	Start-up Time (min)	Mass for 72-hr Mission (kg)	Comments
Stirling[a] JP fuel 1.7 kg	35	21 fuel to electric	20.5		4.6	TRL 4-5. No advanced development. DARPA Palm Power.
Stirling JP fuel 12.6 kg	160	16 fuel to electric	12.7	3	19	TRL 6 demonstrated configured as hybrid source.
IC engine Gasoline 3.1 kg	400	~18 fuel to electric	129	1		TRL 5 integrated into robot in hybrid configuration.
IC engine Gasoline 13.1 kg	1,000	~25 fuel to electric	76.3	1		TRL 9 typical Honda commercial product in limited field use by military
IC engine JP fueled	~1,000					TRL 5-6 Honda engine modified to use JP fuel directly.

Type and Fuel and Dry Weight	Power Level (Watts)	Efficiency (%)	Dry Systems specific Power W/kg	Start-up Time (min)	Mass for 72-hr Mission (kg)	Comments
IC engine JP fueled	~1,000					TRL 4-5 Honda engine operating from reformed JP fuel.

[a] James Huth and Josh Collins, Diesel Fuel-to-Electric Energy Conversion Using Compact, Portable, Stirling Engine-Based Systems," 13th International Stirling Engine Conference, Japan Society of Mechanical Engineers.

In the mid-term (3-5 years), the primary impediment to the further development of Stirling technology for military applications is cost. If sufficient funding is available, mass-produced engines producing 100-500 W and having thermal conversion efficiencies of 30 percent or greater should be available. System specific power in the range of 20-30 W/kg appear feasible. The far term is not clear at this time.

HARVESTED ENERGY

There are many small-scale energy conversion technologies that might form the basis for energy harvesting, but they generally cannot be scaled up to industrial size. Several technologies can be applied to harvesting:

- Piezoelectric materials generate a small voltage whenever they are mechanically deformed. Vibration/pressure from any source can stimulate piezoelectric materials to convert some of that mechanical energy to electrical energy potentially useful to the Soldier. Harvesting of Soldier motion has been demonstrated in the laboratory.
- Some wristwatches are already powered by the kinetic pendulum energy that can be harvested from the movement of a human arm. The conversion mechanism is the movement of a coil in a magnetic field of a permanent magnet, generating usable electrical energy. This is a practical application at low power for watches.
- Photovoltaics_convert the energy in optical radiation, primarily from the Sun, to usable electrical energy using semiconducting materials that are photovoltaic. Many practical applications exist at the Soldier level and are currently employed in theater.
- Thermoelectric generators consist of the junction of two dissimilar materials and the presence of a thermal gradient. Such generators are

highly developed for space applications. They require large thermal gradients at high temperature to attain reasonable conversion efficiency.

- Micro wind turbines convert the kinetic energy in winds to usable electrical energy. The technology in large sizes is commercially available but applications at the individual Soldier level are limited by mass and the need to elevate the turbine.
- In areas of dense radio frequency such as large urban areas, it is possible to use special antennas to harvest the local radio frequency waves across a wide spectral range and convert them into usable energy. Laboratory-scale demonstrations have been conducted, but usable energy requires large antenna structures unless the device can be placed close to a powerful transmitter.

Solar Energy

The solar flux at Earth's surface is on the order of 1,400 W/m^2. At first glance, that constitutes an enormous amount of energy available for harvesting for Soldier use. However, three factors govern our ability to use this energy. First, it only works when the Sun shines and, second, the efficiency of the conversion mechanism determines the amount of solar energy that is harvestable. Thirdly, in current embodiments, it is difficult to use when a Soldier or a squad is on the move.

Moreover, even under the best of conditions, solar energy is available only during daylight hours and is further constrained by local weather and atmospheric conditions. While these factors are limiting, the energy is there for the taking, has minimal exploitable signature, and, in current embodiments, is robust, flexible and lightweight. Theoretical efficiencies can be calculated as a function of the material energy band gap and the junction type for exposure to the solar spectrum.

There is a great disparity between what is theoretically possible and what is achievable in practical embodiments. The state of the art in photovoltaic cells for harvesting solar energy was driven initially by NASA for powering spacecraft and has been applied to operations as far away as Mars, where the insolation is much less than that available at Earth orbit. Since NASA strives to minimize the mass being put into orbit, its research has focused on cell efficiency.

Cost, robustness, and low mass have driven the first field application of photovoltaic energy harvesting to use thin-film amorphous silicon cell technology. The Army currently deploys solar converters that generate up to 60 W and weigh about 10 lb with forces in Iraq and Afghanistan. The solar cells are made from highly flexible, amorphous thin film that can be optimally folded or rolled into a compact package. Amorphous thin-film cells are only about 10 percent efficient and need somewhat less than 0.5 m^2 of active cell area to generate 60 W in full sun conditions.

There are many vendors of this technology. Typical solar converter units that have military potential are available from OkSolar (OkSolar.com), which advertises off-the-shelf systems ranging in power output from 5.0 W to 60 W and varying in weight from 0.17 kg to 1.18 kg for the cell array alone. Under full sun, the 5 W system has a specific energy of 29.4 Whr/kg normalized to 1 hr and the 60 W unit has 51 Whr/kg for the same normalizing condition. Figure I-2 shows a deployable portable solar array.

Near-Term Solar Solutions

Since the solar flux is fixed, improvements in energy conversion efficiency is the only way to increase the energy available to be harvested. Crystalline silicon solar cells have been developed for many years, and the technology is very mature for applications where the cells can be rigidly mounted and where cost is not an issue. Along with the increased efficiency of single-crystal silicon cells comes the fact that these cells can be bent, folded, or otherwise mutilated and still function. However, they cost much more than the amorphous thin-film cells. An increase in efficiency from 10 percent to 25 percent would decrease the area needed for a 60 W charger to about 0.17 m^2. It is still unclear how a solar unit would be packaged for military use and what its performance parameters in finished format would be.

FIGURE I-2. Deployable portable solar array. SOURCE: Deanna Tyler and George Au, Army Power Division, "Assessment of the Army's Need for Portable Energy," presented at the DARPA Limits on Thermodynamic Storage (LOTS) of Energy Proposers' Day Workshop, December 4, 2009.

It is, however, now possible to have 60 W output from a 10 cm × 17 cm (4.0 inches × 6.5 inches) flat panel that could be integrated into the Soldier's outer garments if robustness issues can be overcome. In this way, it could become a source of energy for the Soldier that is always on when the Sun shines and in a hybrid configuration can keep his or her primary energy source, a lithium-based rechargeable battery, in a state of full charge.

Mid- to far-term solar solutions

In the mid term, advances in photovoltaic converter technology will reduce the size of arrays by a factor of two or more, with concomitant reductions in mass. There is considerable room for improvement in the engineering of high-efficiency photocells through multibandgap engineering. GaInP/GaAs/Ge multijunction devices have demonstrated 32 percent efficient cells without concentration. Also, the application of nanotechnology in the form of quantum wells and quantum dots promises further increases in efficiencies, so that a 60 W array will approach the physical dimensions of a playing card, allowing ready integration into the Soldier's outer garments with redundancy.

Biomechanical Energy

Biomechanical energy is harvested by a mechanism that can extract energy from the motion of legs, arms, and other body types to drive a generator that converts some of the kinetic energy into usable electrical energy. For the most part, biomechanical energy harvesting has concentrated on leg motion since legs are used repetitively and repeatedly for locomotion, whereas arm motion is highly variable, and arms often do not move substantially when they are involved in load carrying. Biomechanical energy harvesting is not new; in fact, hand-cranked generators predate the Second World War. There are also commercially available flashlights that are driven by shaking a spring loaded magnetic mass through a linear alternator to produce "minutes" of light. These devices require a conscious effort on the part of the Soldier to harvest energy.

Because Soldiers walk a lot, the challenge is how to harvest energy from this motion. Robert Kunzig has described human locomotion as somewhat like an "imperfect pendulum" (Kunzig, 2001). In a pendulum energetic, energy cycles between potential energy, stored in the vertical lift of the pendulum, and the kinetic kind, as the pendulum passes through the point of lowest potential energy. If walking were ideally pendulum-like, energy use would be minimal, since the center of mass of a walking human passes through the same cycle. The actual walking process is only about 65percent efficient at most, with the remainder of the energy coming from the energy content of the food eaten by the individual. In any case, the energy harvested from locomotion is ultimately provided by energy intake from food.

A survey of the literature shows two efforts that have reached the level of sophistication needed for serious consideration. The first of these, based on initial work at the Simon Frazier University in Canada and further developed and marketed by Bionic Power, headquartered in British Columbia, Canada, harvests energy from knee motion, which is a source of "negative work" as the "pendulum" process involved in walking cycles energy between potential and kinetic. Figure I-3 shows a Bionic Power system intended for military use. By carefully controlling the point in the walking cycle where energy is harvested, the manufacturer claims that its technology has minimal effect on locomotion and fatigue of the wearer.

The performance of the Bionic Power system is as follows:

- Nominal power output: 8-14 W (1.5 m/s walking speed, level ground).
- Maximum power output: 25 W (15 degree down slope).
- Effort setting: 10 levels.
- Output voltage: 5 V to 16.8 V (2 to 4 Li ion cells).
- Maximum output current: 5 A
- Battery chemistries supported: lithium ion (others available upon request).
- LCD indicator: charge complete, charging, fault, output power.
- Connections: left leg, right Leg, battery.
- Fault Protection: Reverse polarity, open/short circuit, over/under voltage, temperature faults.
- External power input: 8 V to 24 V: Solar, vehicle, fuel cell.
- Operational speed: 0.5 to 3 m/s (slow walk to fast jog).

FIGURE I-3. Harvest of biomechanical energy. SOURCE: Reprinted with permission from BionicPower™. Available online at http://bionic-power.com/powerwalk_photos.html.

The second technique was described by Hitt et al.[3] This harvesting is similar in base concept to the harvesting described above in that it takes into account the mechanics of locomotion to harvest energy associated with the rotary motion of the leg-foot interface during the normal walking or running process. Figure I-4a shows the biomechanics of the walking process.

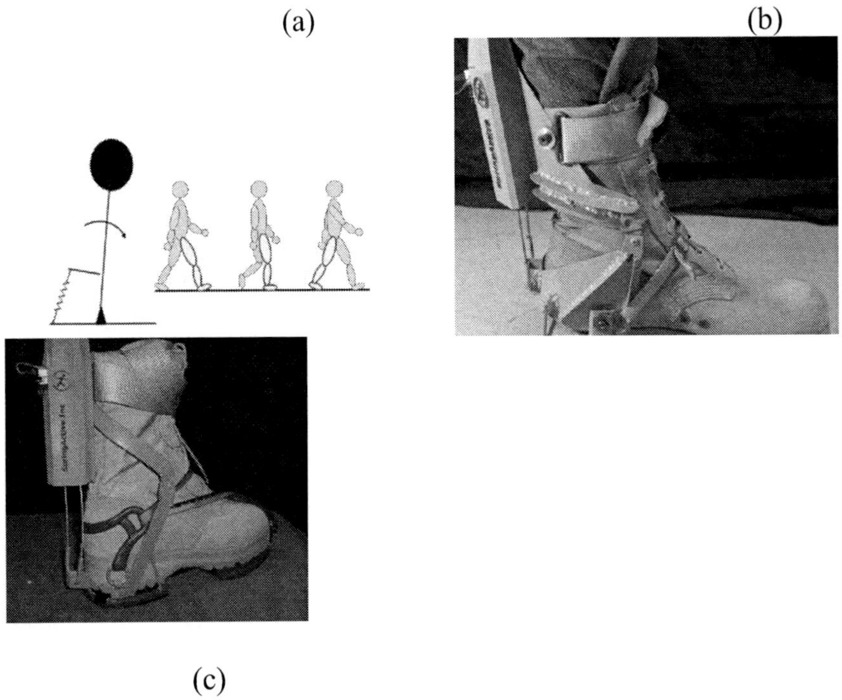

FIGURE I-4 Soldier Power Regeneration Kit (SPaRK): (a) illustration of the biomechanics; (b) side view of device attached to combat boot; and (c) back view of device attached to combat boot. SOURCE: LTC Joseph Hitt, et al., Program Manager, DARPA, "Dismounted Soldier Biomechanical Power Regeneration," presented at the Proceedings of the 27th Army Science Conference, Orlando, Fla., November 29 – December 2, 2010.

The mechanics of the Soldier Power Regeneration Kit (SPaRK) device, which is based on an inverted pendulum model, is described by Hitt et al. as follows.

> Soldier Power Regeneration Kit (SPaRK) will harvest energy during mid-stance of walking gait. The tibia rotates over the stance foot as the contralateral limb swings and positions for heel strike. During this time, the energy in the stance foot ankle joint is negative as the muscles

[3]LTC Joseph Hitt, et al., Program Manager, DARPA, "Dismounted Soldier Biomechanical Power Regeneration," presented at the Proceedings of the 27th Army Science Conference, Orlando, Fla., November 29 – December 2, 2010.

> work to slow the falling body. This mechanical energy was transferred to a small DC motor via a highly efficient ball-screw and burned off across a resistor to determine output power. A uniquely tuned spring was placed in series with the ball screw to reset the starting position during the swing phase. This method allowed the production of 3.5 watts of continuous power output from one ankle device while walking at 6.4 km/hr (4 mph.)

The work by Hitt et al. was intended as proof of principle; typical performance parameters for each ankle are given in Table I-4. The authors conclude that the results provide confidence that biomechanical energy harvesting may be a viable "augmentative and emergency power supply for the future network-centric dismounted Soldier."[4]

TABLE I-4. Energy and Weight of Biomechanical Prototype.

	Speed 4.8	(km/hr) 6.4	Knee Bends
Average power (W)	2.5	3.5	9.2
Average energy/step (J)	2.7	3.2	7.7
Total device weight (kg)	1.4	1.4	1.4
Average W/kg	1.8	2.5	6.6
Generator-only weight (kg)	0.3	0.3	0.3
Average W/kg	8.3	11.7	30.7

SOURCE: LTC Joseph Hitt, et al., Program Manager, DARPA, "Dismounted Soldier Biomechanical Power Regeneration," presented at the Proceedings of the 27th Army Science Conference, Orlando, Fla., November 29 – December 2, 2010.

Biomechanical harvesting of energy from a Soldier's locomotion will continue to mature in the near to mid term, with systems increasing in efficiency and decreasing in mass. This technology will become even more important as parallel efforts to reduce the demand for energy for a Soldier's electronics suite are successful.

Other harvesting mechanisms that have been investigated and may have far-term potential:

[4]LTC Joseph Hitt, et al., Program Manager, DARPA, "Dismounted Soldier Biomechanical Power Regeneration," presented at the Proceedings of the 27th Army Science Conference, Orlando, Fla., November 29 – December 2, 2010.

- Pyroelectric energy harvesting,
- Electrostatic (capacitive) energy harvesting,
- Blood sugar energy harvesting, and
- Tree metabolic energy harvesting.

HYBRID ENERGY TECHNOLOGY

Hybrid systems consist of a primary high-energy-density element, an intermediate rechargeable energy storage unit usually capable of higher specific power, and an energy management system that allows the unit to interface with any load. In most embodiments, the high-energy-density element is a fueled system. Fueled systems derive oxygen for combustion from the air arrangement that has both good and bad features. First, the mass of the oxygen used in the energy production process is not carried by the Soldier and second, the energy content of the fuels is large compared to that of batteries, which both store and deliver energy. On the downside, the fuel's need to ingest air for combustion in turn necessitates the ability to shut the fueled system off when there is a potential for clogging or contaminating the unit. Doing so renders the fueled system unusable until it can "inhale" clean air again. To mitigate these problems, fueled source-rechargeable battery hybrids are being researched, and a few have been introduced on a limited basis to the Soldier. The state of the art for hybrid systems is depicted in Table I-5. The data assume a 72-hr mission.

TABLE I-5. State of the art in hybrid systems

Power Train	Weight (lb/kg)	Volume (in^3/cm^3)	Energy (Whr)	Average Power Capability (W)	Specific Energy (Whr/kg)	Comments
Six Li-145 batteries (all- battery baseline)	13.2/6.0	216/3,540	870	12.0	145	Inventory
4G Zn-air fuel cell/ Li-145 battery	9.4/4.3	218/3,572	1,145	16.0	266	Inventory
Methanol fucl ccll/ Li- 145 battery	6.1/2.8	94/1,540	895	12.4	319	Experimental limited rate initial production

Power Train	Weight (lb/kg)	Volume (in^3/cm^3)	Energy (Whr)	Average Power Capability (W)	Specific Energy (Whr/kg)	Comments
Li-air fuel cell Li-145 battery	NA	NA	NA	NA	NA	Concept

REFERENCES

Kunzig, R. 2001. The physics of . . . walking: Why humans move like an imperfect pendulum. Discover Magazine. Available online http://discovermagazine.com/2001/jul/featphysics. Accessed March 29, 2013.

Laoire, C.O., S. Mukerjee, E. Plichta, M.A. Hendirckson, and K.M. Abraham. 2011. Rechargeable Lithium/TEGDME-LIPF6/O2 Battery. Journal of the Electrochemical Society 158(3): A302-A308.

NRC (National Research Council). 1997. Energy-Efficient Technologies for the Dismounted Soldier. Washington, D.C.: National Academy Press.

NRC. 2004. Meeting the Energy Needs of Future Warriors. Washington, D.C.: The National Academies Press.

Zheng, J.P., R. Y. Liang, M. Henderickson, and E.J. Plichta. 2008. Theoretical Energy Density of Li-Air Batteries. Journal of the Electrochemical Society 155(6): A432-A437.

Appendix J

Lethal and Nonlethal Weapons

This appendix describes current Army weapons programs that are relevant to dismounted Soldiers operating in tactical small units. Only a few of the many items discussed, primarily the individual small arms and ammunition, are contemplated to be fielded to dismounted Soldiers and infantry rifle squads. The larger crew-served mortars and machine guns typically provide support to the squad in platoon operations and are normally organic to weapons squads in the platoon.

LETHAL DIRECT FIRE SQUAD/PLATOON WEAPONS

Weapons in this category pertain to the squad and the platoon. To meet the requirements laid out in LTG Brown's statement on desired lethal effects (see the "Lethality" heading in Chapter 2), improvements must occur in the weapon itself, the ammunition it fires, and the optics and targeting capabilities fielded to the Soldier.

The major improvements to individual and crew-served weapons to meet the lethal requirements include lighter weight, higher reliability in all environments, and higher accuracy at longer ranges. Figure J-1 illustrates the advances made in these areas over the past 10 years and what the future holds for individual categories of weapons. For all these weapons, improvements have reduced the weight, improved reliability, and increased the effective range.

Figure J-2 illustrates the recent and future planned improvements for crew-served weapons. Weight reduction, reliability improvements, and long range accuracy dominate the upgrades to crew-served weapons. Unique to this class of weapons was the elimination of the requirement to set head space and timing on the improved 50 caliber machine gun.

The improvements to individual and crew-served weapons do not end with the programs identified by the Program Executive Office (PEO) Soldier. There is a robust set of programs in the technology base that, if executed, would further reduce the weight and accuracy of these weapons. Figure J-3 is an example of one such improvement program.

In conclusion, capability gaps identified over the past ten years of conflict have led to improvements in individual and crew-served weapons that are lighter weight, more reliable, and more accurate out to the weapons' maximum range. In some cases, the reliability and accuracy improvements have extended the weapons' range beyond their original baseline range.

As noted above, to ensure the Soldier has dominance in lethal effects, improvements are also needed in small caliber and medium caliber ammunition. The most recent example of what can be accomplished is the fielding of the M855A1 enhanced performance 5.56 mm round. Not only is the improved round environmentally friendly, it also has improved performance against hard targets and is extremely effective against a wide variety of targets. With the successful development of this round, the Army intends to continue to improve other calibers in the near future. Unfortunately, much of this small caliber research is unfunded at present, and the projected budget cuts for the Department of Defense (DoD) will only exacerbate the difficulty in implementing these potential improvements.

The third prong in the Army's approach for improving lethality for the dismounted Soldier is to improve optics and targeting technologies. As with the first two elements of the program, improvements in optics and targeting during the past 10 years have enhanced the Soldier's capabilities in combat operations. Figure J-4 illustrates the capabilities either available to the Soldier now or in development in these two areas. There are also programs in the technology base to potentially improve Soldier capabilities in optics and targeting. A few examples of these programs are the Small Arms Smart Sight Equipped Hyper Spectral Camera and the Soldier Wearable Gunfire Detection System Fusion and Networking.

In conclusion, the direct fire capabilities of squads and platoons have improved tremendously over the past 10 years. There are funded programs in both PEO Soldier and the technology base to continue to add to the dominance the U.S. warfighter enjoys in this area. The program that probably best captures the promise of the future is the Lightweight Small Arms Technology Program. The initial phase of the program is well under way with the objective of reducing the weight of the existing M249 machinegun by 35 percent and the weight of its ammunition by 40 percent, while improving lethality, reliability, maintainability, and controllability (through recoil reduction) and keeping the improved version at the equivalent cost of the existing M249.

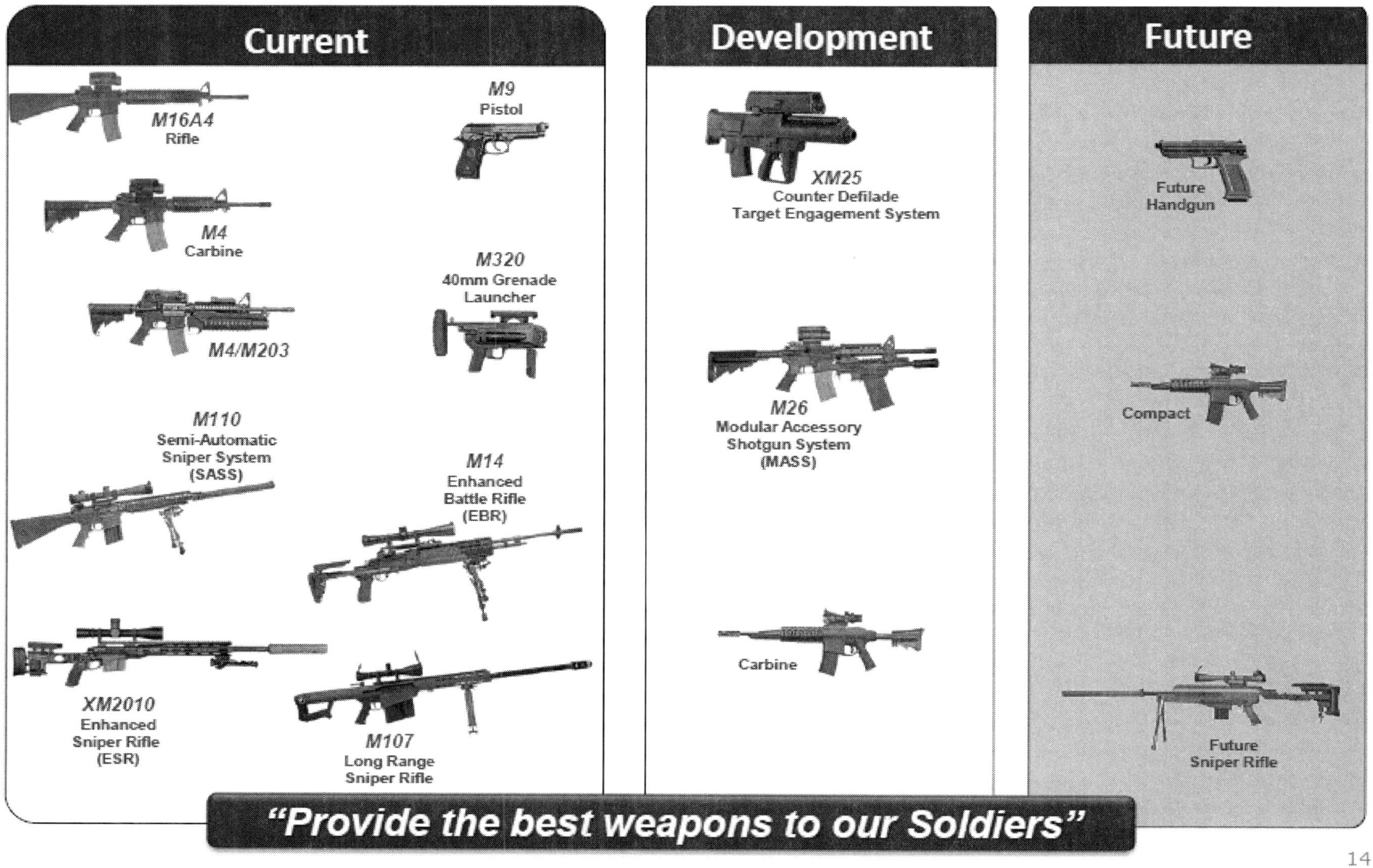

FIGURE J-1. Major improvements to individual and crew served weapons over the past 10 years. SOURCE: Brigadier General Camille Nichols, Program Executive Officer Soldier, "PEO Soldier Overview to the Board on Army Science and Technology (BAST)," presentation to the committee, June 8, 2011.

FIGURE J-2. Current and future crew-served weapons. SOURCE: Brigadier General Camille Nichols, Program Executive Officer Soldier, "PEO Soldier Overview to the Board on Army Science and Technology (BAST)," presentation to the committee, June 8, 2011.

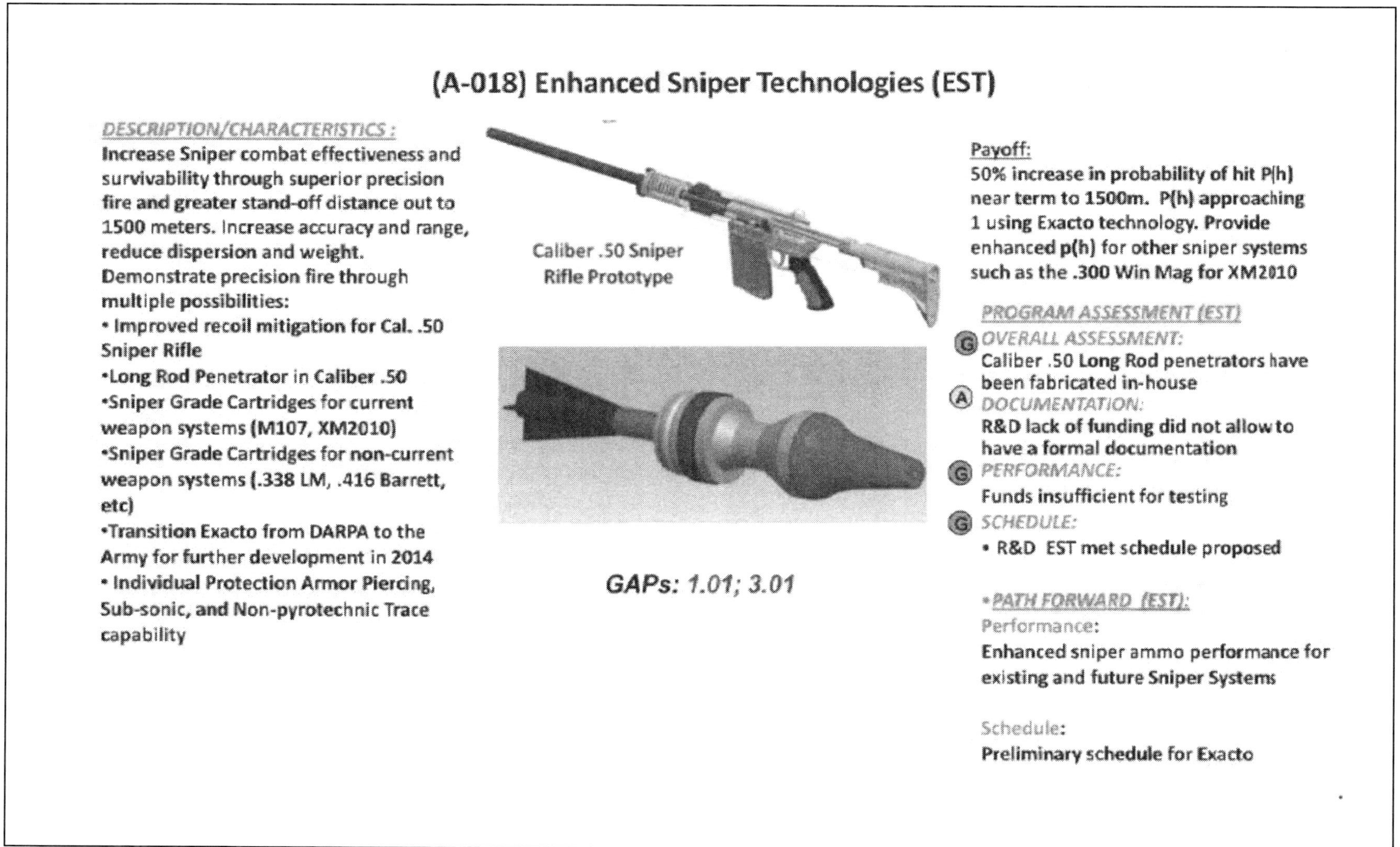

FIGURE J-3. Enhanced sniper technologies. SOURCE: Provided courtesy of the Armament Research, Development and Engineering Center (ARDEC).

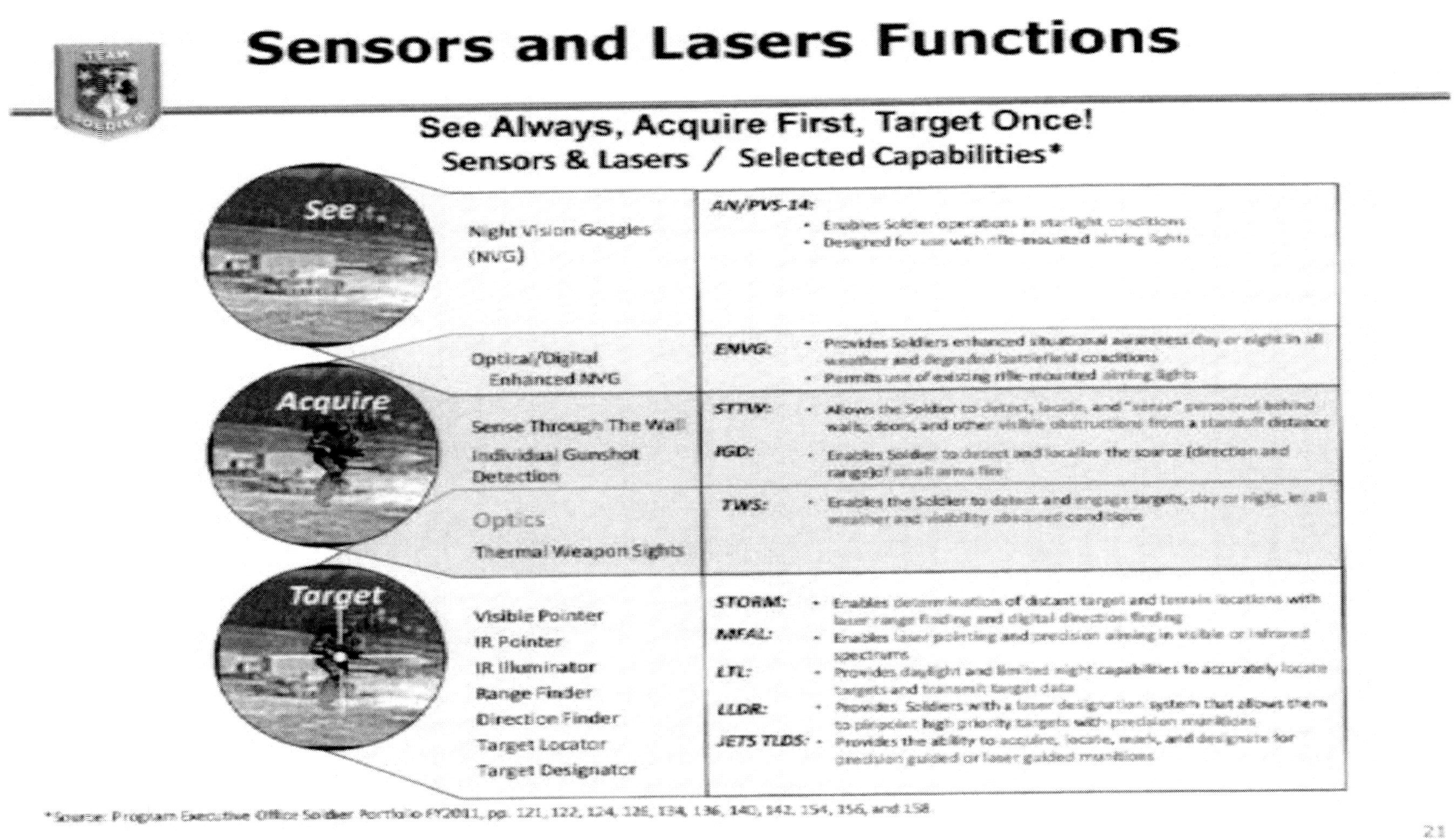

FIGURE J-4. Sensors and lasers functions. SOURCE: Brigadier General Camille Nichols, Program Executive Officer Soldier, "PEO Soldier Overview to the Board on Army Science and Technology (BAST)," presentation to the committee, June 8, 2011.

DEFILADE AND AREA FIRES

Until limited quantities of the new XM25 precision grenade launcher were introduced into the Afghanistan theater for Operation Enduring Freedom, infantry squads lacked the capability to accurately target and defeat enemy combatants protected behind walls or in other defilade positions. The standard grenade launcher and handheld grenades were available, but they do not provide the capabilities of XM25. Nicknamed the "Punisher" by Soldiers, this semiautomatic shoulder-fired weapon has a laser rangefinder and can launch a 25 mm round that explodes at a set distance. It is used to defeat enemy fighters protected by barriers such as walls, rocks, or ditches. Its range is 500 meters for precision targeting and 700 meters for area targets. The following is a quote from a Soldier in Afghanistan:

> If you know anything about Afghanistan you know that the enemy likes to hide behind stuff, and we really can't shoot through boulders and stuff like that. On the first engagement, we were engaged by PKM fire up on the OP. And what happens is you receive fire and you return fire. …
>
> What happened was when we initially received the PKM fire, you reengage with your 240s and your M2s and your M4s, and after we figured we really weren't getting to the enemy enough, [a Soldier] was directed to fire with the XM-25… and like I said before, the enemy likes to hide behind rocks and boulders and we really can't shoot through stuff like that. After [the Soldier] engaged with four rounds of the XM-25, the firefight just ceased. We really couldn't go out and do any BDA or anything like that. But you could tell that when the XM-25 brought the difference to whether they would stay there 15 or 20 minutes shooting, taking pot shots at us—where the actual fight ended after using the XM-25. That was due to the ability of the XM-25 to shoot beyond targets and behind targets.

The Army is currently deciding whether the XM-25 will become a program of record and thus a part of the weapons inventory.

The major area weapon organic to units that provide support to infantry rifle squads is the indirect fire mortar. Figure J-5 illustrates how mortars are distributed throughout the current types of infantry units.

Over the past 10 years, efforts to improve the mortar have focused on lightening components, improving the fire control process, increasing mobility of the 120 mm mortar for light, airborne, air assault, and ranger units, and developing a smart round for the 120 mm mortar. Figures J-6, J-7, and J-8 are examples of these efforts to improve mortars.

Because of the importance of defilade and area fires organic to the infantry squad and their parent organizations in operations conducted in both Iraq and Afghanistan, the Army has dramatically improved capabilities in both areas. The programs to further improve the mortar support the goal of ensuring dismounted tactical small unit (TSU) overmatch in the near and mid terms. Beyond that time frame, there are numerous development programs in the technology base to

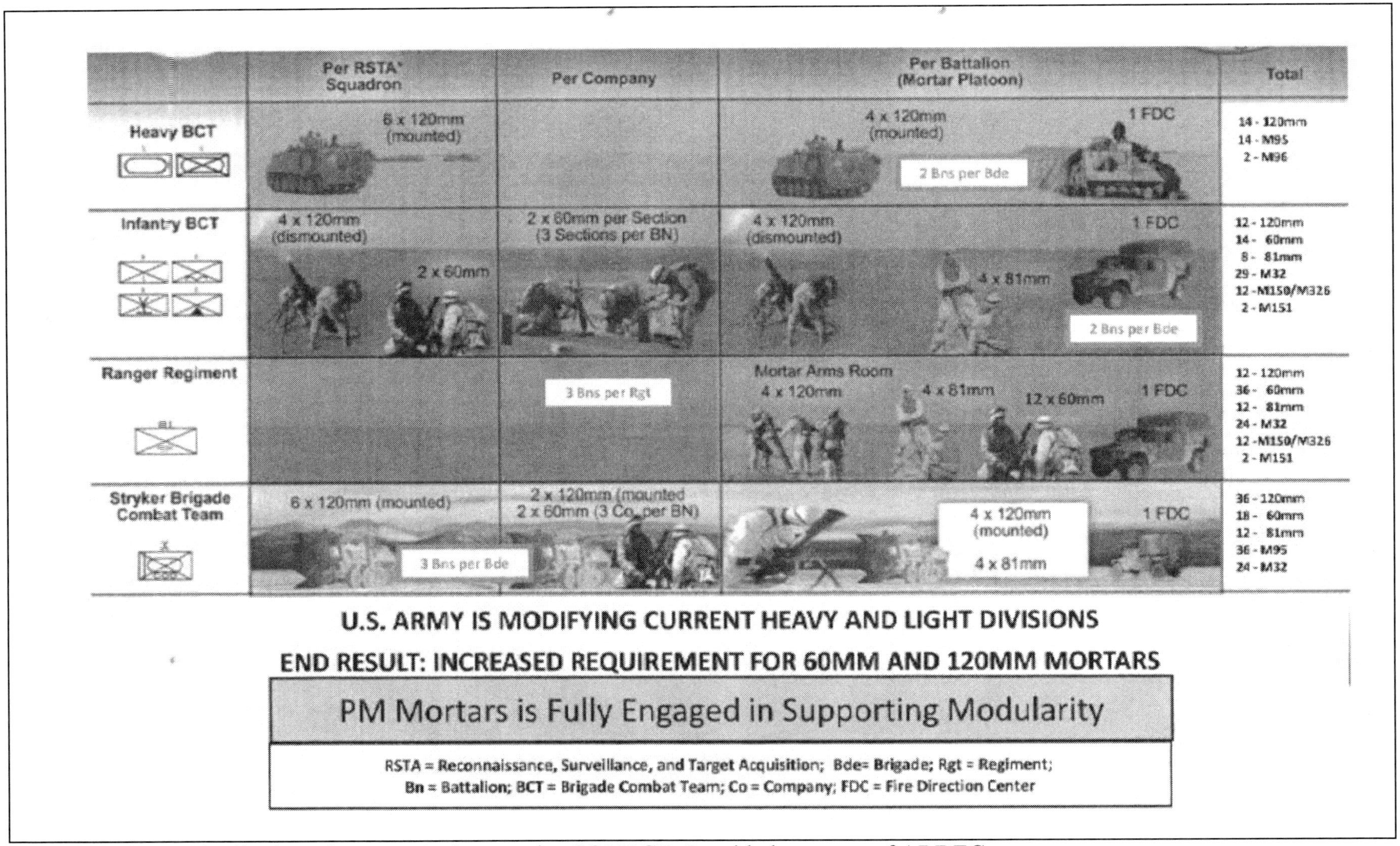

FIGURE J-5. Distribution of mortars in Army units. SOURCE: Provided courtesy of ARDEC.

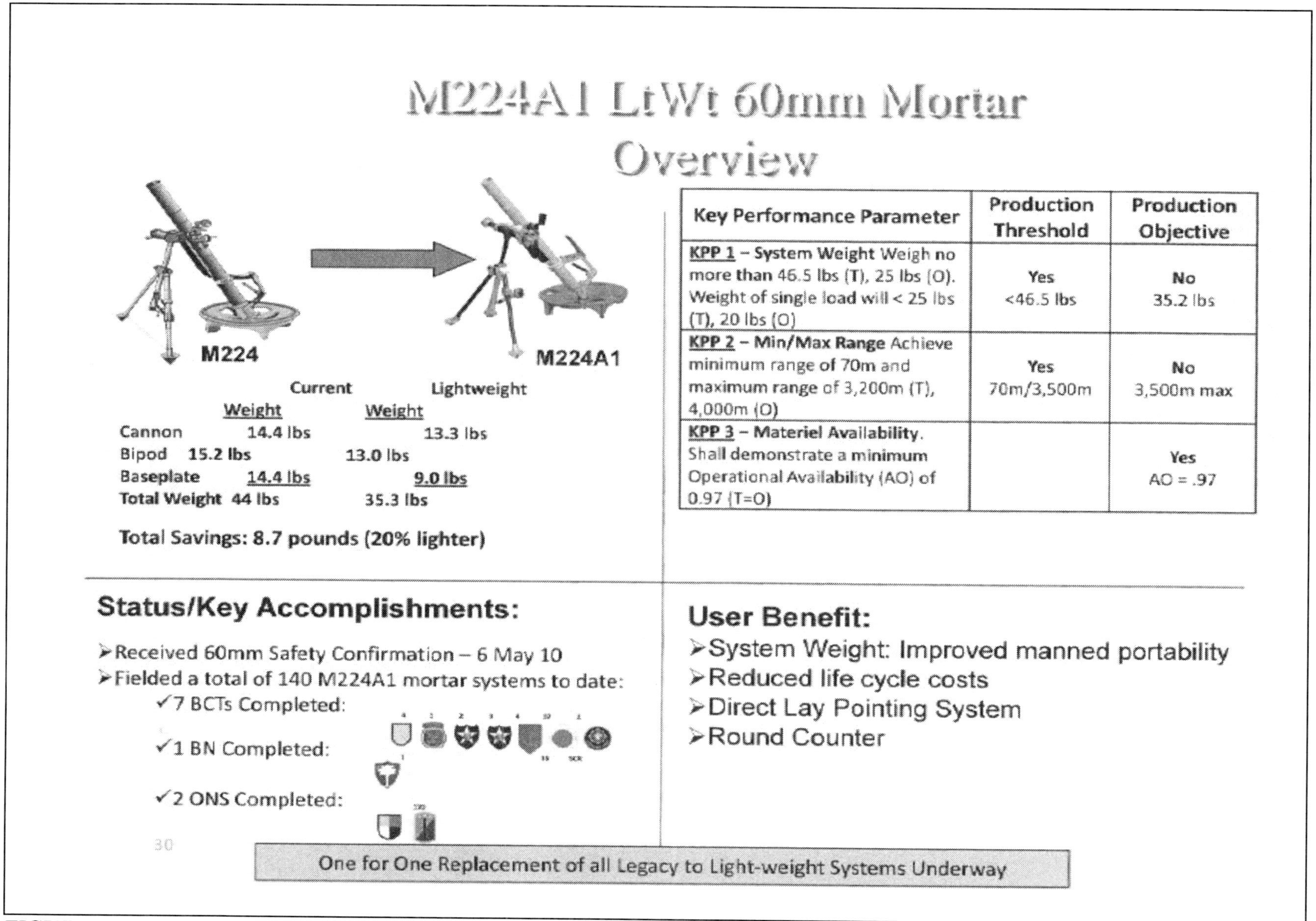

FIGURE J-6. Examples M224A1 Lt Wt 60mm mortars. SOURCE: Provided courtesy of ARDEC.

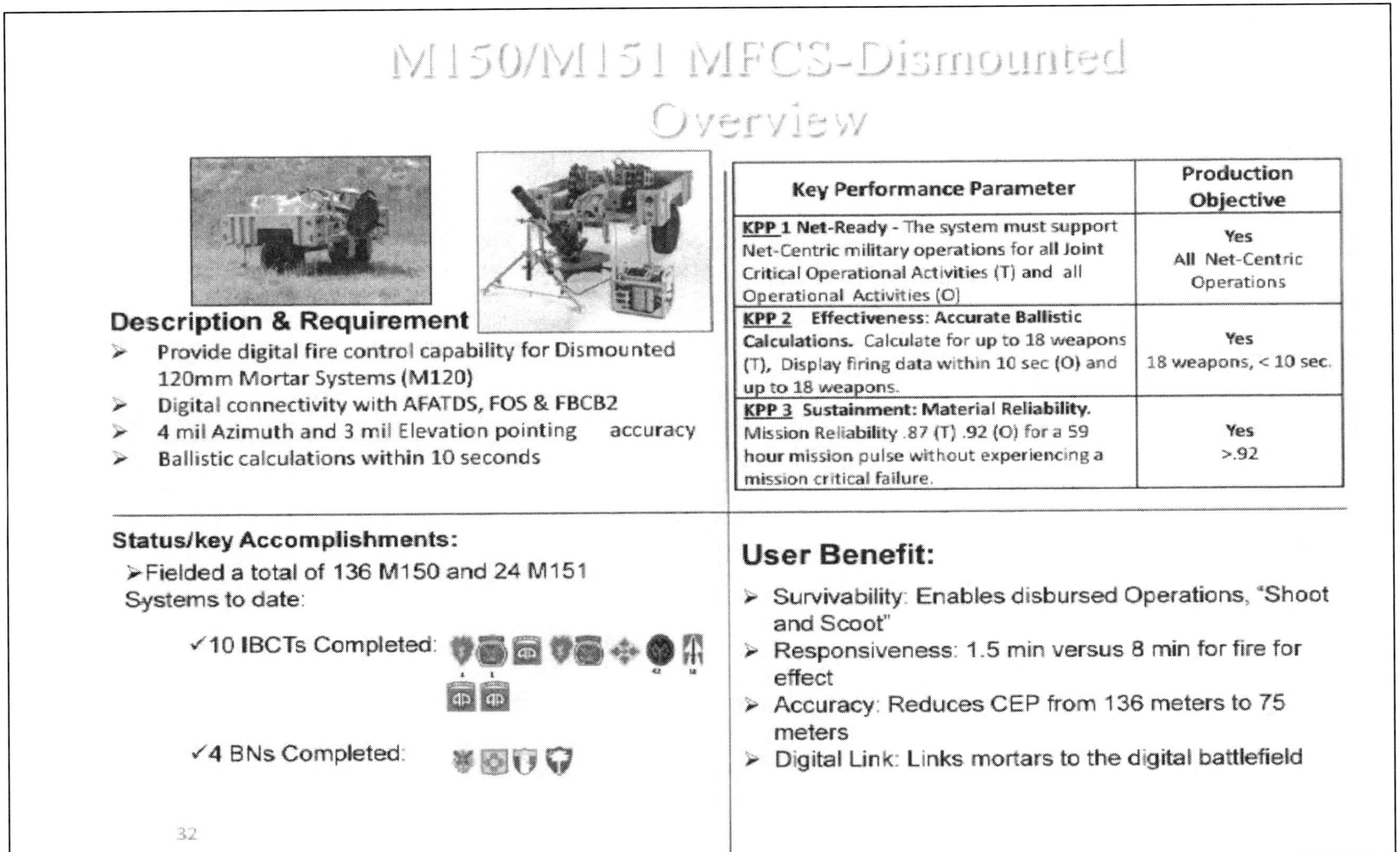

FIGURE J-7 M150/M151 MFCS Dismount. SOURCE: Provided courtesy of ARDEC.

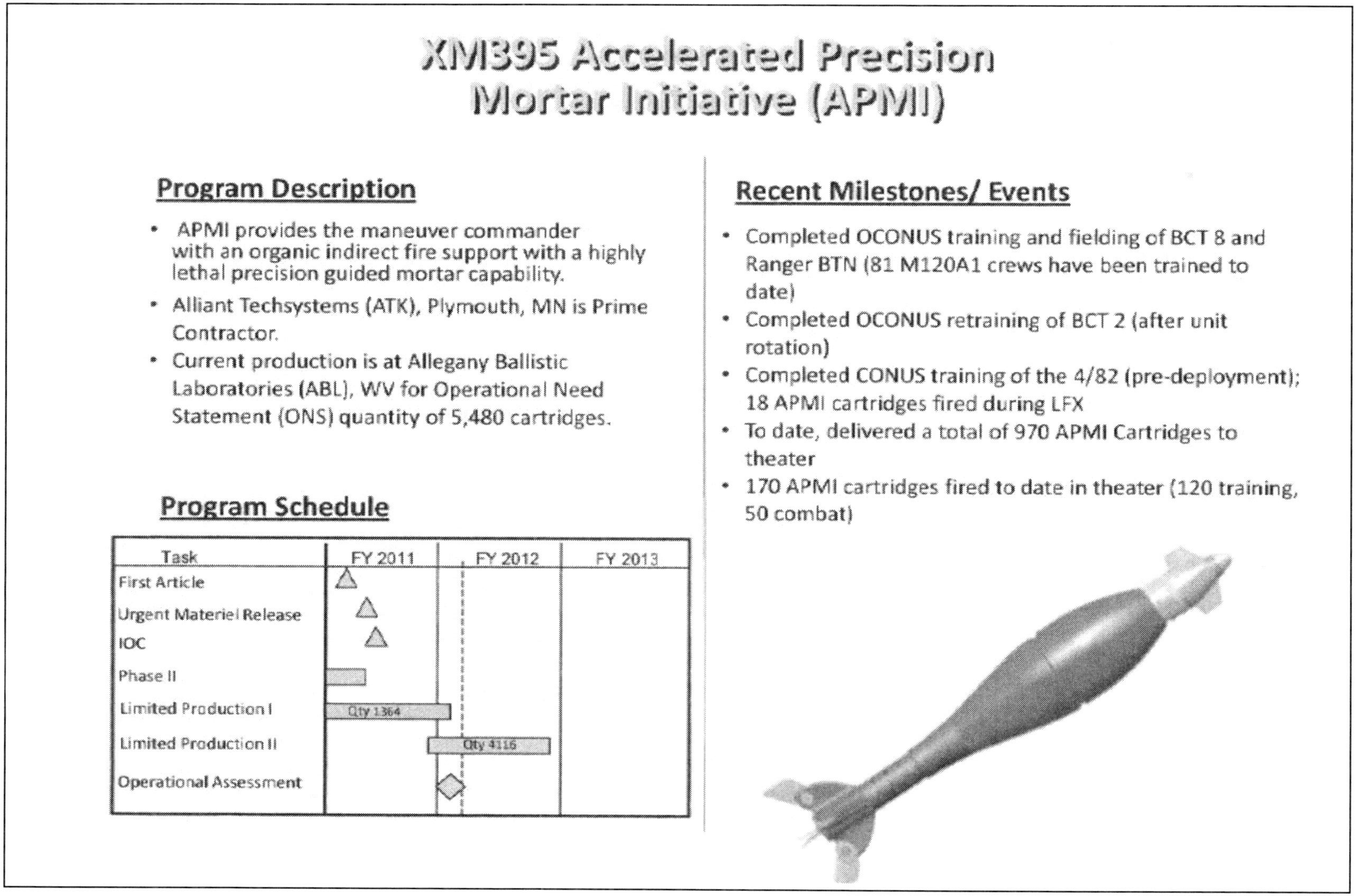

FIGURE J-8. Accelerated precision mortar. SOURCE: Provided courtesy of ARDEC.

expand capabilities in both areas. The Army now has three programs that represent the technology base initiatives in defilade and area fires organic to the infantry squad and their parent organizations:

- Extended range 40 mm guided projectile
- Sensor mortar network
- Precision non-line-of-sight munition technology for light forces

As long as the Army continues operations in Afghanistan, the need to improve the defilade and area fires capability of the infantry Soldier, the dismounted TSU, and supporting units will continue to have a priority. The question is whether this priority will continue, in both programs of record and the technology base, once U.S. forces depart the current areas of operations.

SUMMARY ON LETHALITY FOR THE TSU AND SUPPORTING ECHELONS

The lethality of the infantry Soldier, squad (or future TSU), platoon, company, and infantry battalions of all types have improved dramatically in the past 10 years. If priority and budget remain focused on continuing to improve lethal capability, there are numerous programs in the technology base that can further improve lethality and help ensure the dismounted Soldier remains dominant in all operational environments where lethal force is the determining factor in decisive action.

NONLETHAL WEAPONS

Wikipedia provides the following useful characterization of nonlethal weapons:[1]

> Nonlethal weapons, also called less-lethal weapons, less-than-lethal weapons, non-deadly weapons, compliance weapons, or pain-inducing weapons are weapons intended to be less likely to kill a living target than are conventional weapons.

Nonlethal weapons are used by the Army across the range of military operations involved in unified land operations. Military police, United Nations forces, and occupation forces use them for peacekeeping and stability operations. Nonlethal weapons may also be used to channelize a battlefield or control the movement of civilian populations.

Until the U.S. involvement in Somalia in 1993-1995, the DoD investment in nonlethal capabilities primarily mirrored the capabilities found in civil law enforcement departments and agencies. The material solutions available to Soldiers, Sailors, Marines, and Airmen were developed to support operations such as riot control, crowd control, and self-defense. The

[1] Additional information available at http://en.wikipedia.org/wiki/Non-lethal_weapon.

experience in Somalia led to the establishment of the DoD Non-Lethal Weapons program. The Commandant of the Marine Corps serves as the executive agent of the program. The joint component of the program conducts research on nonlethal technologies. The services' nonlethal organizations also conduct research and are responsible for the development, procurement, and fielding of capabilities.

The tactical challenges in Iraq and Afghanistan have put a premium on the use of nonlethal weapons in both combat situations and their more traditional use in projecting force during law enforcement activities. Initially, capability gaps were met through rapid fielding of singular capabilities. In the spring of 2008, the first multipurpose nonlethal kit was fielded to U.S. Army brigades. The kits came in weatherproof compartments and consisted of five different modules designed for checkpoints, crowd control, detainee operations, convoy support, and dismounted support.

The DoD program consists of multiple service programs to develop or enhance the following nonlethal counter-personnel capabilities:

- Improved flash bang grenade
- Airburst nonlethal munition
- Long range ocular interruption
- Active denial technology
- Improved acoustic hailing device
- Underwater engagement
- MK19 nonlethal munition
- 40 mm human electromscular incapacitation projectile
- Mission payload module
- Nonlethal extended range marking munitions
- Distributed sound and light array
- Vehicle lightweight arresting device
- Single net solution and remote deployment device
- Pre-emplaced electric vehicle stopper
- Small vessel stopping entanglement
- Multi-frequency radio frequency vehicle stopper

Similar to what occurred with lethal capabilities during the past 10 years, the nonlethal area was energized by the operational environments in both Iraq and Afghanistan. The individual infantry Soldier and his units have benefited greatly from the numerous nonlethal material capabilities introduced in both theaters. The joint program and the individual services have developmental programs to further add to current capabilities. As is also the case for lethality improvements, the future of improvements in Soldier nonlethal capabilities lies in the ability of DoD to continue to fund the numerous developmental programs robustly during the near, mid, and far terms.